No. 903
$9.95

DO-IT-YOURSELFER'S GUIDE TO MODERN ENERGY-EFFICIENT HEATING & COOLING SYSTEMS

BY JOHN TRAISTER

TAB BOOKS
Blue Ridge Summit, Pa. 17214

FIRST EDITION

FIRST PRINTING—DECEMBER 1977

Printed in the United States
of America

Library of Congress Cataloging in Publication Data

Traister, John E
Do-it-yourselfer's guide to modern energy-efficient heating & cooling systems.

Includes index.
1. Dwelling—Heating and ventilation—Amateurs' manuals. 2. Dwellings—Air conditioning—Amateurs' manuals. 3. Dwellings—Energy conservation—Amateurs' manuals. I. Title.
TH7225.T7 697 77.20854
ISBN 0-8306-7903-0
ISBN 0-8306-6903-5 pbk.

Preface

This book is aimed at you, one of this country's fast-growing contingent of do-it-yourselfers. Perhaps you want to economize. Perhaps you can't get repairmen at any price to do a first-class job. Or perhaps money is no object and your repairmen do excellent work, but you still want to do the repairs around your home yourself—mainly because you enjoy it.

If you fit into any of the categories above, or if you plan to build, buy, or remodel a home, you will find this comprehensive guidebook helpful. It gives all the information you need to know on how to choose, use, maintain and even repair your heating and cooling system for year-round comfort.

You'll learn how to estimate the heating and cooling capacity needed for a single room or an entire home; how to avoid expensive repair bills; and how to live comfortably for less and get the most out of your utility dollar. It's all here in this book, a book no home or home owner should be without.

I wish to thank the many manufacturers of heating and cooling equipment for their help in supplying me with reference materials and the many illustrations in this book.

John E. Traister

Contents

Chapter 1
Preparing Your Home For Heating & Cooling

Few items in your home cost more to operate than your heating and cooling system. Year after year billions in costly fuels are consumed by these systems to provide comfort for home owners and their families. But did you know that much of this energy escapes unused because many homes are not properly prepared for using these systems? Besides the enormous waste in energy and dollars, additional money is needlessly spent by the home owner for larger-than-necessary equipment to offset those wasted Btu's.

We are all aware of the spiraling cost of fuel and the need to reduce our energy consumption, but many of us don't exactly know how to reduce this wasted energy. Look no further! This chapter will show you how to immediately reduce your wasted energy as much as 30%–60%, without taking up too much of your time or money.

SEALING CRACKS AND CREVICES

Cracks and crevices inevitably appear in every home over a period of time, even a well-built one. Beside allowing outside air into the home and conditioned air out, moisture, insects, dust, and dirt also enter. Therefore, the sealing of these cracks and crevices should be one of your first projects in improving your home.

Fig. 1-1. Cartridges of caulking compound can be quickly slipped into a top-loading caulking gun.

Caulking

Caulking compound, the "magic" material that clings to wood, stone, or metal, is probably all you will need unless the cracks are over 1/2-inch wide; then use oakum or mortar first. Cartridges of caulking compound with built-in nozzles can be slipped quickly into a top-loading caulking gun (Fig. 1-1), and disposal of empty cartridges is very easy.

In general, you will need to caulk around frames of windows and doors, between foundation and siding, around pipes where they penetrate walls (like an outside water faucet), and especially where a wooden structure joins masonry walls. A putty knife or similar tool should be used to pry away old compound or rot before applying the new caulking.

Caulking should be an annual job, even if your home is relatively new; this way, the job will be minor each year and you will keep your fuel bills at a minimum.

While we're on the subject of sealing cracks, you should also check your window panes every year or so. If you find some of the putty needs replacing, remove the old putty with an old chisel or glazier's knife, and provide an airtight seal around the pane with a thin bed of glazing compound or putty. To apply, string the putty between thumb and forefinger (Fig. 1-2), wiping it into the rabbets in which the pane fits. Then use the tip of a putty knife to smooth it.

Weatherstripping

Weatherstripping is equally as important as caulking for saving fuel and insuring your family's comfort in winter. The cracks around windows and doors of an average house—if not weatherstripped—leak as much air as a three-foot-square hole in your exterior wall.

Weatherstripping comes in several different types: felt, wood and felt, metal, and metal and felt. All of these will do the job, but the all-metal types are usually the easiest to apply.

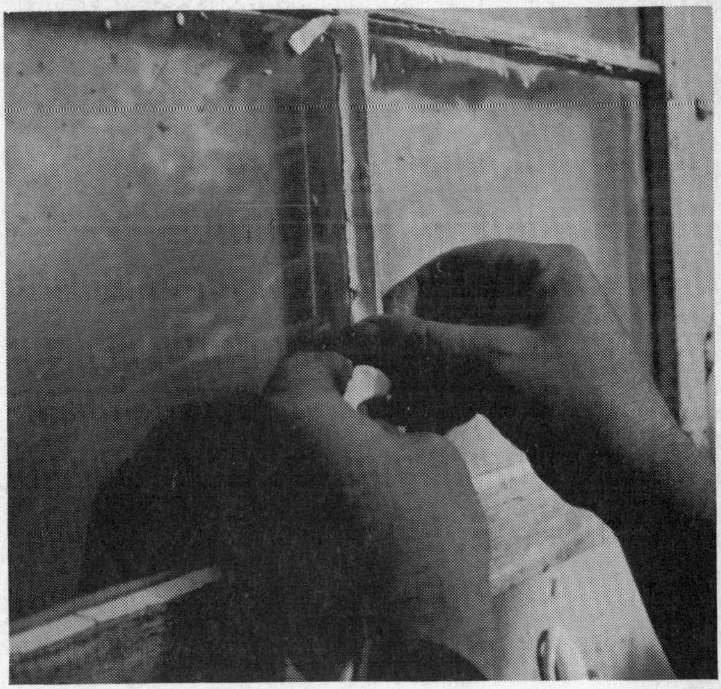

Fig. 1-2. Method of applying putty to seal window panes.

Simply reel the stripping out of its container, cut it to fit with scissors, and nail it in place (the nails are usually included).

Metal strips can be installed on windows by opening the window completely and inserting the weatherstripping in the track where the window sash runs. Install the strip so it barely clears the window stop when the window is shut, compressing the springiness of the metal strip. Nail stripping to the sill also, to seal the gap under the sash when the window is closed.

All exterior doors should be weatherstripped in similar fashion. However, the door bottom is often sealed with special metal-and-felt weatherstripping instead of ordinary types. The direction of opening and the clearance between door and threshold determines whether this piece is screwed to the inside or outside of the door.

INSULATION

Insulation is certainly not new: man has used natural substances for centuries to keep heat in or out of his living quarters. The rough thatched huts of northern Europe and the British Isles had thick straw-covered roofs, while huts in Africa and the South Seas were (and are) covered with hollow sea grass and reeds for insulation.

Today, building experts agree that insulation is vital to holding down heating and cooling bills, and its importance to realty values is recognized by all housing authorities.

To save heat loss in winter and cut down on heat gain in summer, insulate the attic, floor, and walls with the proper amount of quality insulation.

Attic spaces in both new and existing homes should be insulated with a minimum of four inches of insulation. Six inches is better, and some home owners even use batts of fiberglass insulation 10 inches thick, especially when the home is electrically heated. If the attic space is open and readily accessible, fiberglass batts or loose insulation poured from a bag may be used. However, many homes have flooring in the attic which prevents using batts or poured insulation. In this case blown insulation is the answer. Removing a few floorboards and using an insulation blower will enable you to fill the space between the ceiling and the attic floor with insulation.

Most insulation dealers will loan or rent you a small portable blower when you buy insulation. They'll also provide you with complete instructions on its use. Be sure to wear a respirator when installing insulation—it's bad stuff for lungs. If you prefer to have the work done by others, look in the classified section of your telephone directory under "Insulation Contractors."

New homes are normally insulated by insulation board on the outside of the studs (between the studs and the outside finish) and four-inch-thick fiberglass batts between the studs. In existing homes with wood siding, one of the top siding boards is removed around the perimeter of the house, and loose insulation is either poured or blown into the space between the studs. Alternatively, small holes can be drilled through the siding at the top of each between-stud cavity to accommodate the nozzle of the blowing machine. Then the siding is carefully replaced or wooden plugs are used to plug the holes before the surface is refinished.

If the house has masonry outside the studs, there are still several ways in which the side walls can be insulated; the method depends on the structural characteristics of the house. Some home owners drill down through the partition plate in their attics and then blow insulation down into the partitions through these holes. However, since most homes have bridging or fire stops installed about halfway down between studs, the blown insulation will not completely fill the partition. In order to fill this space, the baseboard along the outside walls can be removed, holes drilled in the walls behind the baseboard, and insulation blown through these holes. You may be able to blow the insulation up from the basement, and since the loose insulation is somewhat compressed as it is being blown in, very little will fall back down. Of course, all holes should be plugged and sealed after the job is finished.

Fiberglass batts are best for insulating under floors. Purchase the type with paper backing and of the correct width; the batts can be secured to the floor joists with a staple gun. Figures 1-3, 1-4, and 1-5 show various types of house construction and how they should be insulated.

The cost of insulating your home will very quickly be returned in fuel savings. From then on, it's money in the bank (your account, not the fuel company's). Living comfort plus the increased resale value of your home after insulating are additional returns on your investment.

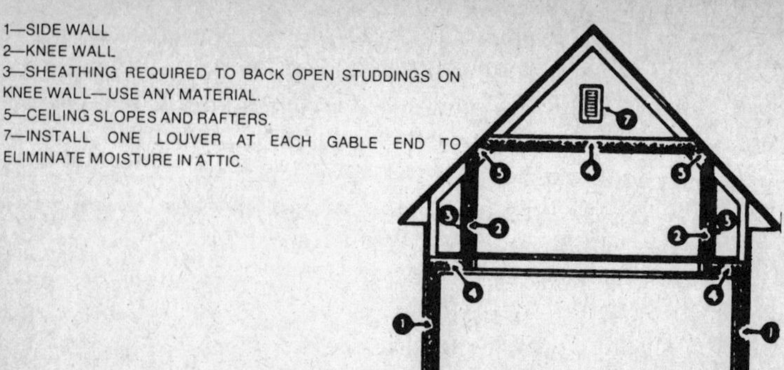

Fig. 1-3. Insulating a home with a second-floor knee wall.

ADDING STORM WINDOWS AND DOORS

Most experts agree—and many studies have shown—that most of a home's heat loss occurs through an uninsulated roof. However, recent studies using infrared cameras suggest that loss through doors and windows may be equally as significant.

Sealing cracks around frames of windows and doors, as mentioned earlier, will help reduce heat loss, but the addition of storm windows and doors or double glass will do wonders in cutting your fuel bills. In fact, some homes' fuel bills have been cut in half by a complete storm-window-and-door treatment.

Double glazing—two sheets of glass separated by a dead air space—is most easily incorporated in a new home, but if you don't mind the expense, you may order double-glazed frames to fit in your present window openings. It's far less expensive to add conventional storm windows, though. If your

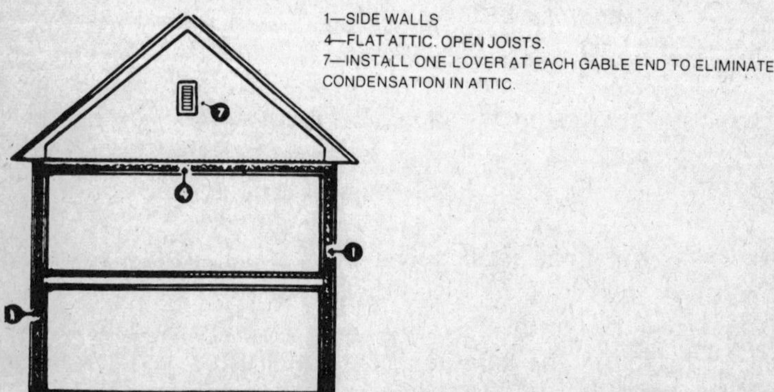

Fig. 1-4. Insulating a home with vented, unheated attic.

windows are a non-standard size, storm windows can be quickly and inexpensively ($10-18 per window) made to fit.

Should you want to do something immediately about the heat loss through your windows, you can purchase inexpensive snap-in clear plastic storm windows that are installed on the inside of your windows. No special tools are needed as the self-adhesive backing holds the mounting trim to the window frame. Most of these come in a kit which includes a plastic sheet, vinyl mounting trim for silled windows, and easy-to-follow instructions. If the standard sizes will not fit your windows, the trim may be cut with a regular handsaw and the plastic shield with a pointed utility tool. You can install these on every window in your home in less than a day.

To add conventional storm/screen insulating windows, measure the window opening from the outside: width is the distance between side casings; height is the distance from the window sill to the top of the casing. Allow a sixteenth of an inch or a little more clearance to compensate for out-of-square frames and to ensure easy installation.

When installing, make sure you don't distort the shape of the frame: the sides must remain straight and parallel. Some windows come with spring clips which hold the frame rigidly against the storm insert; the clips should not be removed until installation is complete. If your windows don't have such clips, check the fit of the insert by moving it up and down in the frame before you secure the second side of the frame.

Most windows are predrilled to accept small woodscrews. The screws will attach to the window casing more easily if you drill starter holes, using the holes in the storm frame as a

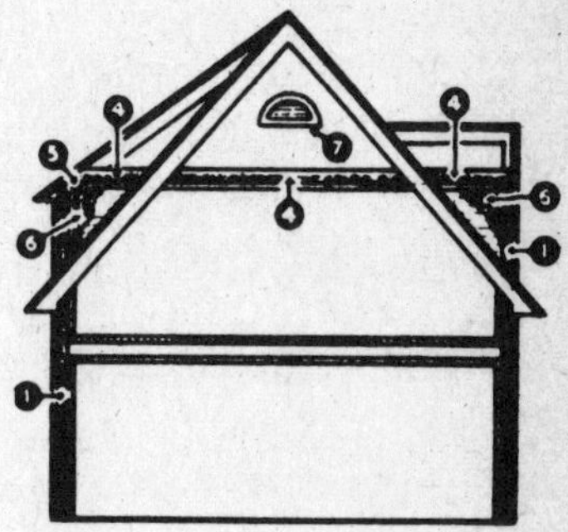

Fig. 1-5. Method of insulating a home with dormers.

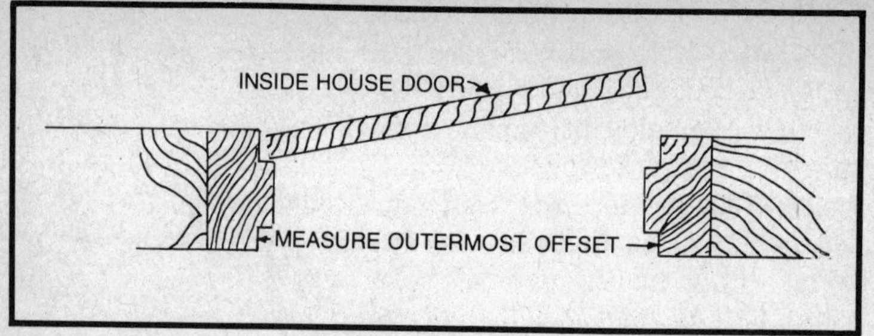

Fig. 1-6. Measuring a door opening to determine size of storm door.

guide. If the frame holes are badly positioned for the window, drill new frame holes—the soft aluminum frame is easily drilled with a standard twist bit.

Storm doors are installed similarly to the windows, but the door size should be measured as follows. Face the door opening from the outside and measure the distance between the facing boards at the outermost offset (as shown in Fig. 1-6). Take measurements at the extreme top and bottom. Then measure the height of the door from the bottom of the top facing board or brick molding to the top of the threshold; this measurement should be made at both sides. Use the smaller of the two width and height measurements when ordering the new door.

REPAIRING LOOSE-FITTING SIDING

The sealing of minor structural cracks was mentioned earlier and can be accomplished using caulking compound. However, wood siding works loose over the years too; this siding should be repaired to keep the heat loss and gain to a minimum.

Renail all loose siding on your home and replace any that is rotten or damaged. If there are still slight gaps, use caulking compound or some other type of sealer on them. This sealing will not only cut down your fuel bills, but you will also find that your home is easier to keep clean, since less dust and dirt will enter.

HUMIDITY CONTROL

Now that your home is insulated and sealed up tight, your heating and cooling bill should drop considerably, but there is one problem that you are going to encounter if you don't take precautions—excessive moisture. Efficient insulation can create a moisture problem because all the cracks and leaks that allowed heat to escape in an uninsulated home also carried away the excess moisture. With these cracks filled, the moisture remains inside your home.

There are, however, several solutions to the problem of excessive moisture, the best being the use of vapor barriers which limit the moisture flow through your walls. In new construction, vapor barriers (usually sheets of polyethylene) should be installed as follows:

1. All walls and floors exposed to outside temperatures should have a separately applied vapor barrier.
2. The vapor barrier should be located on the warm or heated side of the insulation; this applies to all surfaces.
3. The vapor barrier should be secured to a firm support at all openings such as windows, doors, electrical outlets, etc.
4. A vapor barrier should be installed over dirt floors of crawl spaces and basements.
5. A vapor barrier should be installed under all on-grade and below-grade slab floors.
6. All joints should overlap at least three inches.

In an existing home where it is impossible to install a separate vapor barrier, it is recommended that paints that retard the flow of moisture be used on the inside room surfaces. Reputable dealers can show you figures illustrating the relative ability of various brands of paint to resist moisture penetration.

There are also mechanical devices that will help provide your home with moisture control. Exhaust fans—like the ones used in your bathroom or kitchen—can be used to control moisture. The fans should be vented to the outside of the structure through the roof or wall and equipped with an automatic damper.

The fans should be controlled by an automatic humidistat connected in parallel with the manual fan switch. When so connected, the user may turn the fan on or off at any time when the humidistat is not calling for the removal of moisture. However, if the humidistat is set for, say, 35–40% relative humdity and moisture content becomes greater than this amount, the fan will operate automatically, overriding the manual control.

During summer months, the humidistat should be set at 100% relative humidity or at the OFF position in order to prevent the fan from running continuously. Of course, you may also purchase a dehumidifier for removing excessive moisture, but if the recommendations given in the preceding paragraphs are followed, this should not be necessary.

MINIMIZING SOLAR LOADS

Minimizing solar loads in summer by shading glass areas and ventilating attic or roof cavities will reduce your cooling requirements tremendously. Not only will you save on fuel, but you will need smaller, less expensive equipment.

Attic and crawl spaces should be ventilated if the insulation is to perform at its best. Vents should never be closed as ventilation is just as important in winter as in summer. The insulation will retard the flow of heat while the vents let unwanted moisture escape. The following are suggestions for planning proper ventilation:

1. At least two vent openings should be provided for each closed area. Attic vent openings should have a minimum net free area of one square foot of inlet and one square foot of outlet for every 300 square feet of ceiling area. Insect screens will reduce the otherwise-free area by 50%.
2. Crawl space vent openings should have a minimum net free area of one square foot of inlet and one square foot of outlet for each 1500 square feet of crawl space.

Ventilating fans installed in the attic can lower attic temperatures 30 degrees or more, resulting in a much cooler house during the hot summer months and substantial savings in air-conditioning bills.

There are several ways to ventilate your attic. Power vents with automatic thermostats are available for either gable-, or roof-mounting, in sizes from 2000 to over 5000 CFM. Or you may want to install a turbine ventilator—one that costs nothing to operate! The slightest breeze starts the airfoil blades moving to exhaust hot air from attic spaces. When installing any of these fans, make certain you have sufficient intake openings to provide make-up air.

Large areas of glass let a lot of the sun's heat into your home—during summer months, this can raise your cooling bill

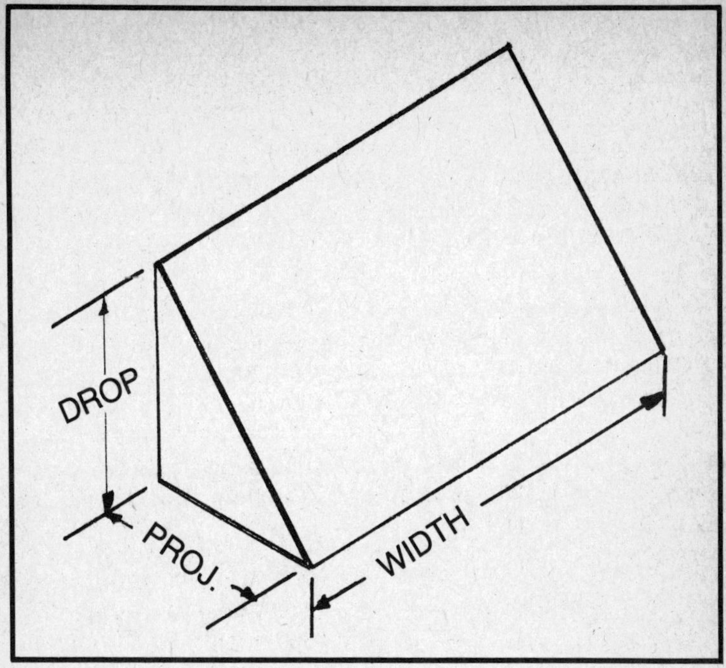

DROP
PROJ.
WIDTH

Fig. 1-7. Measurements needed for ordering and mounting awnings.

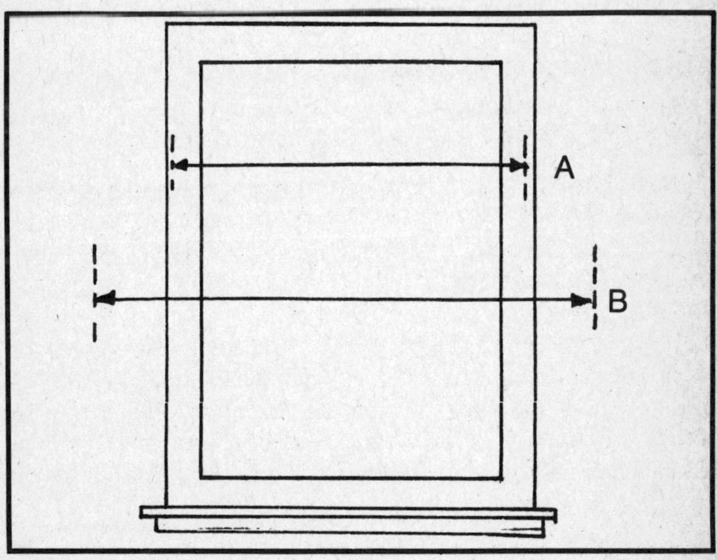

Fig. 1-8. Possible mounting positions for awnings: (A) on frame, (B) on siding.

considerably. One quick solution is to install a reflective sun-control film over picture windows and other large glass areas. With the kits available, you can do the job yourself in only minutes for about $.80 per square foot. Such a shield also reduces heat loss, and, for that matter, heat gain through the glass during the winter.

Awnings are another means of reducing the effect of the sun's heat. The old short-lived canvas types are now obsolete, having been replaced with attractive, durable aluminum. To measure your windows for awnings, refer to Fig. 1-7 to see how to obtain drop, projection and width. Figure 1-8 shows possible awning mounting positions: on the window frame (A) or on the siding (B).

Other fuel-saving techniques include the planting of shade trees and the use of appropriate colors for your roof and walls. Remember that a light color reflects heat while darker colors absorb it.

Chapter 2
Simplified Heating Calculations

One of the requirements for comfort during cold weather is maintaining the home at a predetermined temperature and humidity often considerably different from that of the outside air. Therefore, it is necessary to provide a heat source to meet this requirement.

A good heating system features:

1. Adequate, dependable, and trouble-free operation
2. Comfort during cold weather
3. Reasonable installation cost
4. Reasonable annual operating cost
5. Ease of service and maintenance

In order to best realize these goals, there are certain steps which must be taken to determine the best type of system to use, the size of the equipment, and the best fuel. However, before any of these factors can be determined, a complete survey of the building's structural conditions must be made along with heat loss calcuations. It is with the latter that this chapter will deal.

Heat loss is expressed in either Btu per hour or in watts. Both are measures of the rate at which heat is transferred and may be converted from one to the other by the two formulas which follow:

$$\text{Watts} = \frac{\text{Btu}}{3.4} \qquad \text{Btu} = \text{Watts} \times 3.4$$

Basically, determining heat loss through walls, roof, ceiling, windows, and floor requires just four steps:

1. Determine the net area in square feet.
2. Find the proper heat-loss factor from tables.
3. Allow for infiltration.
4. Multiply the area by the factor; the product will be expressed in Btu.

Heat loss calculations can be made more quickly and efficiently using a prepared form such as the one shown in Fig. 2-1. With spaces provided for all necessary data and calculations—in the proper order—the procedure becomes routine and simple. All that's required is that you feed the spaces on the form the correct information and make certain that your calculations are accurate. The result is a

HEATING CALCULATION FORM

NAME OF PROJECT OR AREA________________________

DESIGN CONDITIONS: Outside Dry Bulb Temperature______

Inside Dry Bulb Temperature______ Temperature Difference______

PART 1: Ceiling Type and Insulation Thickness______________

Floor Type and Insulation______________________

Type of Windows________________________________

Type of Walls and Insulation___________________

PART 2: Infiltration

$$\frac{\text{area width} \times \text{area length} \times \text{ceiling height}}{60}$$

$$\frac{(\quad)\text{width} \times (\quad)\text{length} \times (\quad)\text{ceiling height}}{60}$$

	Heat Loss
=__________ x 1.08 x ______T.D. =	________

PART 3: Heat Loss

		Heat Loss
Windows:	______sq. ft. x 0.55 x ______T.D. =	________
Walls:	______sq. ft. x 0.13 x ______T.D. =	________
Roof:	______sq. ft. x 0.20 x ______T.D. =	________
Floor:	______sq. ft. x 0.09 x ______T.D. =	________
TOTAL HEAT LOSS IN BTUH		________
Divide by 3.4 to obtain heat loss in watts		________

Fig. 2-1. Heat-loss estimating form.

room-by-room survey showing exactly how much heat will be required to maintain your home at a comfortable temperature during cold weather.

You will notice that the form in Fig. 2-1 requires exact dimensions of windows, doors, ceilings, floors, and the like. While you're taking these measurements is also a good time to make a floor plan of your home. Such a sketch will not only make identification of particular areas easier when you are using the calculation form, it will also help to lay out the system components. So before we get into the actual calculations, let's briefly discuss how to lay out the floor plan.

Floor plans are not made full size because the size of the drawing would be impractically large. Therefore, the drawing is reduced so that all the distances on the drawing are smaller than the actual dimensions of your home, but all dimensions are reduced in the same proportion. The ratio, or relation between the size of the drawing and the size of the building, is indicated on the drawing (1/8 inch = 1 foot, for example), and the dimensions shown are the actual building dimensions, not the distance that is measured on the drawing.

The most common method of reducing dimensions in proportion is to choose a certain (small) distance and let that distance represent 1 foot in the building. This distance is then divided into 12 parts; each one of these parts represents one inch. If half-inch divisions are required, these twelfths are further subdivided. When measurement is laid off on the drawing, it is made with a reduced foot rule calibrated as described above, called an *architect's scale*. When a measurement is taken on the building itself, it is made with a standard rule or tape measure.

Architect's scales are available in many different ratios: you can draw your floor plan in scales ranging from 1/16 inch = 1 foot to 1 1/2 inches = 1 foot. The scale 1/4 inch = 1 foot is the most common scale for residential drawings. You can even buy drawing paper ruled in 1/4-inch squares to simplify the drawing.

PREPARING THE SKETCH

If your home was designed by an architect, chances are you have a complete set of working drawings already. If not, check with the architect for copies of the drawings. The contractor who built your home will probably have a set you can use for the heat loss calculations. If all these sources fail to produce a set, you'll have to take measurements and draw your own.

Begin your sketch by having a friend help you measure the outside perimeter of your home, drawing the wall lines and inserting dimensions as you go. Then measure and locate all windows and doors on the drawing along with chimneys, carports, etc. Continue by measuring all inside partitions, doors, etc., and indicating their location on your sketch. When you have finished, your drawing should look something like Fig. 2-2 with dimensions added.

You should also note where you have storm windows and doors; the amount of insulation in your attic, outside walls, and under your floor; the type of roof, floor, etc. The blank spaces on the calculation form in Fig. 2-1 will serve as a guide as to what information is required for a complete heat-loss estimate. Try to obtain all this information before you begin your calculations; you will save time and come up with a more accurate estimate.

COMPLETING THE HEAT-LOSS CALCULATIONS

The heat-loss calculation form is based on *design conditions* of the home and the area it's located in. *Outside design conditions* are the extremes of temperature occurring in a specific locality, while the *inside design condition* is the degree of temperature and humidity that will give optimum comfort.

The *outside dry-bulb temperatures* for calculating heating loads are shown in Fig. 5-2 in Chapter 5. For cities and localities not listed in this table, use the design temperatures of listed cities that most closely approximate your local conditions.

The *inside dry bulb temperature* for heating in colder months is 70 to 75°F. Therefore, the *temperature difference* will be the difference between the inside design temperature (say 70°F) and the outside design conditions for your locality. For example, let's assume that you live in Washington, D.C.; referring to Fig. 5-2, we find that the outside design condition for winter heating is 0°F. So the temperature difference is 70 degrees (70° − 0° = 70°). However, if you lived in Miami, you would have a temperature difference of only 35°F, since the outside design temperature for Miami (see Fig. 5-2) is 35°F.

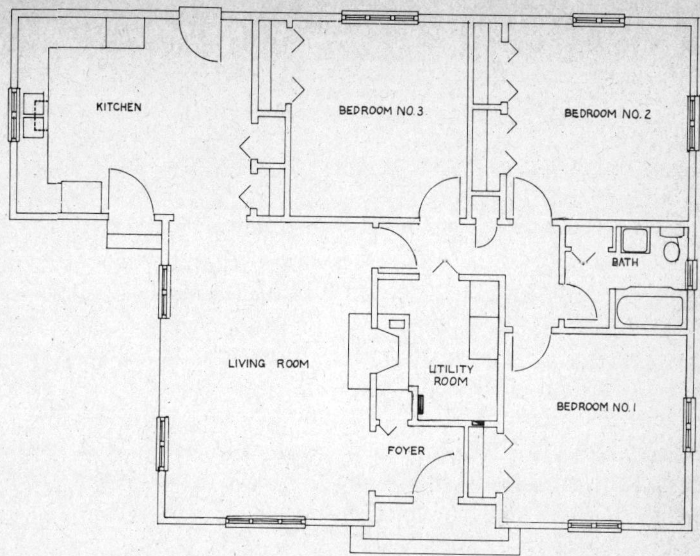

Fig. 2-2. A typical residential floor plan.

In order to demonstrate exactly how to make heat loss calculations for a given home, let's use the floor plan of the residence in Fig. 2-2 as an example. Furthermore, we will assume that this home is located near Washington, D.C., and use design conditions for this area. So, with the inside and outside design conditions and the drawing of the house floor plan, we are ready to do a complete room-by-room heat-loss calculation for the house. We'll use a separate form for each room in the house.

We begin by filling in Part I of the living room form. This information includes:

1. Ceiling below ventilated attic: 2 inches of insulation
2. Floor with crawl space: 2 inches of insulation
3. Windows: single pane with drapes
4. Walls and insulation: plasterboard with 3 1/2 inches of insulation

Part II of the form covers heat loss as a result of *infiltration*, or air leaks through cracks in walls and around windows and doors. Our example living room is 13.75 feet wide by 19.25 feet and has an eight-foot ceiling. When these dimensions are inserted in their proper position in the blanks, we can complete the formula as follows:

$$\frac{13.75 \times 19.25 \times 8}{60} = 35.29$$

Part III of the form includes radiant and conductive heat losses through exterior surfaces, the first being the windows. Therefore we now find the area of all windows in the living room and enter the answer in the appropriate space on the calculation form. We find that the total area of the windows is 49 square feet.

Item No. 2 in Part III of the form is outside walls: one of the living room walls is 19.25 feet in length while the remaining outside wall is 13.75 feet in length. Since we know the ceiling height is 8 feet, the gross outside wall area may be found by:

19.25 feet + 13.75 feet × 8 feet = 264 square feet.

However, the calculation form calls for the net wall area (gross wall area – window area) so we must subtract 49 square feet from 264 square feet to obtain the answer: 215 square feet. Again, this figure is entered in the appropriate space on the form.

The remaining two spaces are for the areas of the floor and ceiling and since these two areas will have the same dimension, we can complete the calculations for both in one step:

19.25 feet × 13.75 feet = 264.68 square feet (of floor and ceiling)

This completes all the data that is necessary for the calculation of the heat loss which must be taken from the drawings; the remaining factors are printed on the form.

Now begin with the first figure (infiltration), multiply by the infiltration factor (1.08) and the temperature differential (7°) and enter the product in the appropriate space. Continue down the list until heat loss is calculated for all areas listed. Summing our subtotals, we find the room heat loss is 11,884.6 Btu/h.

This means that 11,884.6 Btu/h will be required to keep the inside temperature at 70°F if the outside temperature is 0°F. If electric heat is being used and we must convert the heat loss in

HEATING CALCULATION FORM

NAME OF PROJECT OR AREA Living Room

DESIGN CONDITIONS: Outside Dry Bulb Temperature 0°

Inside Dry Bulb Temperature 70° Temperature Difference 70°F

PART 1: Ceiling Type and Insulation Thickness 2″

Floor Type and Insulation Crawl Space 2″ Insulation

Type of Windows Single Pane

Type of Walls and Insulation 3 1/2″ Insulation

PART 2: Infiltration

$$\frac{\text{area width} \times \text{area length} \times \text{ceiling height}}{60}$$

$$\frac{(13.75)\text{width} \times (19.25)\text{length} \times (\ 8\ )\text{ceiling height}}{60}$$

= 35.29 x 1.08 x 70 T.D. = Heat Loss 2667.70

PART 3: Heat Loss

				Heat Loss
Windows:	49 sq. ft. x 0.55 x	70	T.D. =	1886.5
Walls:	215 sq. ft. x 0.13 x	70	T.D. =	1956.5
Roof:	264.68 sq. ft. x 0.20 x	70	T.D. =	3705.1
Floor:	264.68 sq. ft. x 0.09 x	70	T.D. =	1668.80

TOTAL HEAT LOSS IN BTUH 11,884.6

Divide by 3.4 to obtain heat loss in watts 3,495.5

Fig. 2-3. Completed heat-loss form for the living room of the residence in Fig. 2-2.

Btu/h to watts. Divide 11,884.6 by 3.4 and we find 3495.5 watts of electric heat is required to do the job.

The completed heat-loss form for the living room appears as Fig. 2-3. Briefly review this entire procedure before continuing on to the next area.

Of course, if you were adding a heating unit to just this room, and heating the remaining areas with the existing heating system, you could stop now. This method of calculating heat loss is good for one small area or an entire home.

The kitchen/dining area, bedroom, and bathroom calculations are performed exactly as for the living area, using different dimensions. The results of the calculations are

HEATING CALCULATION FORM

NAME OF PROJECT OR AREA Bedroom No. 1

DESIGN CONDITIONS: Outside Dry Bulb Temperature 0°

Inside Dry Bulb Temperature 70° Temperature Difference 70°F

PART 1: Ceiling Type and Insulation Thickness 2″

Floor Type and Insulation 2″

Type of Windows Single Pane

Type of Walls and Insulation 3 1/2″

PART 2: Infiltration

$$\frac{\text{area width} \times \text{area length} \times \text{ceiling height}}{60}$$

$$\frac{(12)\text{width} \times (12.5)\text{length} \times (8)\text{ceiling height}}{60}$$

= 20 x 1.08 x 70 T.D. = Heat Loss 1512.0

PART 3: Heat Loss

Windows:	28 sq. ft.	x 0.55 x	70 T.D.	=	1078.0
Walls:	164 sq. ft.	x 0.13 x	70 T.D.	=	1492.4
Roof:	144 sq. ft.	x 0.20 x	70 T.D.	=	2016.0
Floor:	144 sq. ft.	x 0.09 x	70 T.D.	=	907.2

TOTAL HEAT LOSS IN BTUH 7005.6

Divide by 3.4 to obtain heat loss in watts 2060.47

Fig. 2-4. Completed heat-loss form for bedroom No. 1.

shown in Figs. 2-4 through 2.7. However, the procedures for calculating the utility room is somewhat different because there is no outside window or outside walls; obviously these two steps are omitted when calculating the heat loss for this area. The calculated results for this area are shown in Fig. 2-8.

The heating equipment selected for these areas must meet or exceed the calculated heat loss which, in turn, will maintain the required temperature inside the building. However, heating equipment for any individual room should not exceed the calculated loss by more than 10 percent if at all possible. In duct systems, the duct losses should be calculated before determining the required heat output. A rule of the thumb for duct loss is to multiply the calculated heat loss by a factor of 1.15—1.25.

HEATING CALCULATION FORM

NAME OF PROJECT OR AREA Bedroom No. 2

DESIGN CONDITIONS: Outside Dry Bulb Temperature 0°

Inside Dry Bulb Temperature 70° Temperature Difference 70°

PART 1: Ceiling Type and Insulation Thickness 2″

Floor Type and Insulation 2″

Type of Windows Single Pane

Type of Walls and Insulation 3 1/2″

PART 2: Infiltration

$\frac{\text{area width x area length x ceiling height}}{60}$

$\frac{(12.25)\text{width x }(15.5)\text{length x }(8)\text{ceiling height}}{60}$

	Heat Loss
= 20.41 x 1.08 x 70 T.D. =	1542.8

PART 3: Heat Loss

Windows:	28 sq. ft. x 0.55 x 70 T.D. =	1078.0
Walls:	172 sq. ft. x 0.13 x 70 T.D. =	1565.2
Roof:	156 sq. ft. x 0.20 x 70 T.D. =	2184.0
Floor:	156 sq. ft. x 0.09 x 70 T.D. =	982.8
TOTAL HEAT LOSS IN BTUH		7352 8
Divide by 3.4 to obtain heat loss in watts		2162.58

Fig. 2-5. Completed heat-loss form for bedroom No. 2.

HEATING CALCULATION FORM

NAME OF PROJECT OR AREA Bedroom No. 3

DESIGN CONDITIONS: Outside Dry Bulb Temperature 0°

Inside Dry Bulb Temperature 70° Temperature Difference 70°F

PART 1: Ceiling Type and Insulation Thickness 2″

Floor Type and Insulation

Type of Windows Single Pane

Type of Walls and Insulation 3 1/2″

PART 2: Infiltration

$\frac{\text{area width x area length x ceiling height}}{60}$

$\frac{(12)\text{width x }(12.5)\text{length x }(8)\text{ceiling height}}{60}$

	Heat Loss
= 20 x 1.08 x 70 T.D. =	1512.0

PART 3: Heat Loss

Windows:	21 sq. ft. x 0.55 x 70 T.D. =	808.5
Walls:	75 sq. ft. x 0.13 x 70 T.D. =	682.5
Roof:	156 sq. ft. x 0.20 x 70 T.D. =	2184.0
Floor:	156 sq. ft. x 0.09 x 70 T.D. =	982.8
TOTAL HEAT LOSS IN BTUH		6169.8
Divide by 3.4 to obtain heat loss in watts		1814.69

Fig. 2-6. Completed heat-loss form for bedroom No. 3.

HEATING CALCULATION FORM

NAME OF PROJECT OR AREA Bath

DESIGN CONDITIONS: Outside Dry Bulb Temperature 0°

Inside Dry Bulb Temperature 70° Temperature Difference 70°F

PART 1: Ceiling Type and Insulation Thickness 2″

Floor Type and Insulation 2″

Type of Windows Single Pane

Type of Walls and Insulation 3 1/2″

PART 2: Infiltration

$\frac{\text{area width x area length x ceiling height}}{60}$

$\frac{(6.5)\text{width x }(8.25)\text{length x }(8)\text{ceiling height}}{60}$

= 7.14 x 1.08 x 70 T.D. = Heat Loss 539.7

PART 3: Heat Loss

Windows: 4 sq. ft. x 0.55 x 70 T.D. = 154.0

Walls: 48 sq. ft. x 0.13 x 70 T.D. = 436.8

Roof: 52 sq. ft. x 0.20 x 70 T.D. = 728.0

Floor: 52 sq. ft. x 0.09 x 70 T.D. = 327.6

TOTAL HEAT LOSS IN BTUH 2186.1

Divide by 3.4 to obtain heat loss in watts 642.97

Fig. 2-7. Completed heat-loss form for the bathroom.

HEATING CALCULATION FORM

NAME OF PROJECT OR AREA Utility Room

DESIGN CONDITIONS: Outside Dry Bulb Temperature 0°

Inside Dry Bulb Temperature 70° Temperature Difference 70°F

PART 1: Ceiling Type and Insulation Thickness 2″

Floor Type and Insulation 2″

Type of Windows Single Pane

Type of Walls and Insulation 3 1/2″

PART 2: Infiltration

$\frac{\text{area width x area length x ceiling height}}{60}$

$\frac{(5.5)\text{width x }(9)\text{length x }(8)\text{ceiling height}}{60}$

= 6.6 x 1.08 x ___ T.D. = Heat Loss 498.4

PART 3: Heat Loss

Windows: ___ sq. ft. x 0.55 x ___ T.D. = ___

Walls: ___ sq. ft. x 0.13 x ___ T.D. = ___

Roof: 49.5 sq. ft. x 0.20 x 70 T.D. = 693.0

Floor: 49.5 sq. ft. x 0.09 x 70 T.D. = 311.5

TOTAL HEAT LOSS IN BTUH 1502.9

Divide by 3.4 to obtain heat loss in watts 442.02

Fig. 2-8. Completed heat-loss form for the utility room.

Chapter 3

Selecting Heating Equipment

The importance of selecting the right heating system for your home cannot be overemphasized. Whether you are building, buying, or merely renovating your present home, the type of heating plant you install is all-important. There are several reasons for this.

A properly designed heating system is one of the greatest comforts and conveniences any home can offer. Besides automatically maintaining a comfortable temperature inside your home when it's cold outside, a heating system can provide the correct amount of moisture in the air; remove dirt, bacteria, smoke, pollen, and other impurities; and provide continually circulating fresh air.

However, there are other factors you must consider besides comfort. The way your family lives is one thing that should be considered. How much time is actually spent at home? How much unoccupied space is in your home? Does one member of your household—like an elderly relative—like the temperature in his or her room to be at least 78 ° F, while the rest of your family prefers 70 ° F? How many bedrooms are vacant most of the time? All of these things are factors to consider when selecting a heating system for your home.

To some extent, the construction features of your home will dictate the type of heating system you can use. Other questions you might consider are:

- Does the house have a basement?
- Is it feasible to install ductwork in the existing structure, or would this type of installation require too much cutting and patching?
- Can you afford the best, or should you start with a fundamental system to which you can add later?
- Will you need to maintain a constant temperature twenty-four hours a day, or will you need a system that can provide quick heat when you arrive home in the evenings?

Then you'll want to consider which type of fuel to use. This will mostly be determined by where you live, as the distance from your home to fuel sources affects the fuel cost. But don't always judge fuel by its initial cost. A fuel that costs more initially, but provides clean and automatically controlled heat, could be the cheapest in the long run. This may not always be the case, but is at least worth considering.

The important thing is to make up your mind what you want from your heating system. Then consider the type of building, its location, and the fuel you want to use. If you can swing the best at this time, then what you probably want is a central comfort-conditioning system with automatic dampers in the ductwork to provide different temperature zones in your home; an automatic humidifier to provide the correct amount of moisture in the air; and a electrostatic air filter that provides fresh, germ-and-pollen-free air. You will probably add an air-conditioning system at the same time for summer cooling, so your system will provide all of the above features twenty-four hours a day the year round.

If you would like to have all the above someday, but find it too expensive to purchase all at one time, there is still a way to reach your goal. Start off with an efficient, economical, central heating plant with its related ductwork. Then, in a year or two, begin adding additional components to the system until you have one that offers maximum comfort. The installation of these components will probably cost a little more when added later, but this is offset by spreading the cash outlay over a longer period of time.

Your first addition will probably be cooling components. Chapter 7 of this book shows how you can install these components yourself—in a weekend. An electrostatic filter that will provide clean, pure air should be your second choice, while the humidifier will probably come next. Of course, you do not have to add the components in this order; your own requirements and desires will determine your priorities. Also, don't hesitate to seek the advice of others. A local mechanical contractor, architect, engineer, even a neighbor may be able to offer advice that will help you obtain a better heating system at the lowest cost.

The forced-air central heating system is certainly not the only type available. There are many different types of residential heating systems, and another type may suit your requirements better.

HOT WATER BASEBOARD HEATING SYSTEM

A zone hydronic (hot water) heating system permits selection of different temperatures in each zone of the home. Baseboard heaters located along the outer walls of rooms provide a blanket of warmth from floor to ceiling; the heating unit also supplies domestic hot water simultaneously, through separate circuits. A special attachment coupled to the hot-water unit can be used to melt snow and ice on walkways and driveways in winter and a similar attachment can be used to heat your swimming pool during the spring and fall seasons.

A typical hot-water system operating diagram is shown in Fig. 3-1, and is explained as follows. When a zone thermostat calls for heat, the appropriate zone valve motor begins to run, opening the valve slowly; when the valve is fully opened, the valve motor stops. At that time, the operating relay in the hydrostat is energized, closing contacts to the burner and the circulator circuits. The high-limit control contacts (a safety device) are normally closed so the burner will now fire and operate. If the boiler water temperature exceeds the high-limit setting, the high-limit contacts will open and the burner will stop, but the circulator will continue to run as long as the thermostat continues to call for heat. If the call for heat continues, the resultant drop in boiler water temperature—below the high limit setting—will bring the burner back on. Thus, the burner will cycle until the thermostat is satisfied; then both the burner and circulator will shut off.

Hot water boilers for the home are normally manufactured for use with oil, gas, or electricity. While a zoned hot-water system is comparatively costly to install, the cost is still competitive with the better hot-air systems. The chief

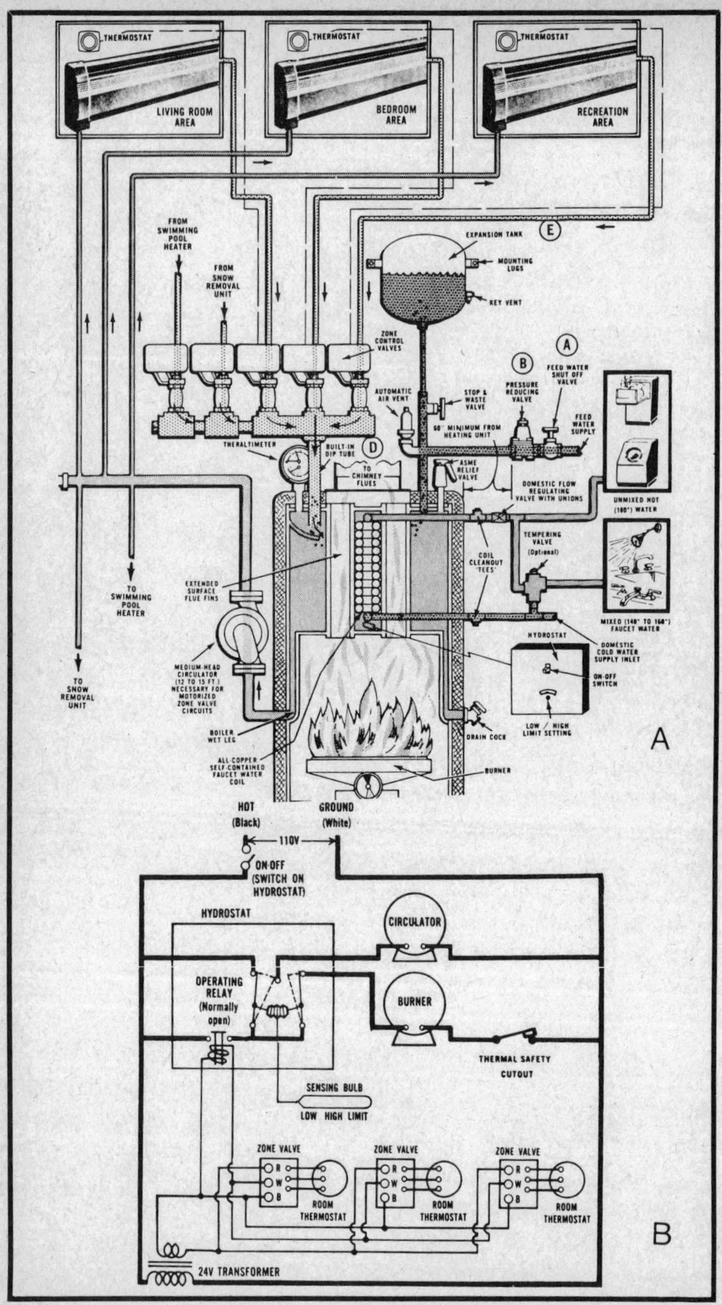

Fig. 3-1. (A) A typical hot-water system operating diagram. (B) Typical electrical wiring.

disadvantage of hot-water systems is that they don't use ducts. If you wish to install central air conditioning, you must install a complete duct system along with the central unit.

ELECTRIC BASEBOARD HEATERS

The heating of homes with electric heat was almost unheard of two decades ago. Now, however, it shows signs of becoming one of the principal types of heat for modern residences. This is due to thc following advantages of electric heat:

1. Electric heat doesn't involve combustion and therefore, in some ways, is safer than combustible fuels.
2. It requires no storage space, fuel tanks, or chimneys.
3. It requires considerably less maintenance than other types of heating systems.
4. The initial installation cost is low.
5. Each room may be controlled separately by its own thermostat.

There are two definite disadvantages to using electric baseboard heaters in the home: (1) humidification is hard to control; (2) the cost of heating with electricity, in most areas, is more than with other fuels under similar conditions.

Still, many home owners find the electric baseboard heaters fit their heating requirements well. Suppose you want to replace your old coal-fired gravity hot-air system with a more modern type, say, an oil-fired forced-air system. Your chimney flue may be adequate for the new oil burner and the existing coal bin may be large enough to house the fuel tank, but what about the new ductwork? A survey of your home shows that much cutting of structural members and cutting and patching of finished walls will be required to run the new ductwork, especially to the second floor. In fact, so much cutting and patching that this phase of the installation will cost nearly as much as the heating system itself.

Then you consider electric baseboard heat. The units can be fastened directly to the wooden baseboard along the outside walls; feeder and control wiring can be "fished" from the units to the electric panel through wall paritions with a minimal amount of cutting and patching; the second floor of your home can be turned off when it's not being used; and the initial cost of installing the electric baseboard units will be considerably less than a forced-air system.

You would have to offset the higher cost of electricity by insulating better and installing storm windows and doors. But these procedures should be a part of any heating system—regardless of the fuel—anyway.

Even in homes with other types of heating systems, you may still find a use for electric baseboard units. For example, let's assume you are adding a recreation room in the basement of your home which has a forced-air heating system. In order to heat the new area, you made a tap in your main duct line and ran an air supply into the basement. Although the system keeps the basement warm enough, the area is still uncomfortable due to the cold, damp floor. What do you do now? Install a couple of electric baseboard heaters along the wall/floor line. This way your forced-air system will take care of most of the heat loss in the area, while the electric baseboard heaters will help maintain an even temperature throughout the room besides warming up the basement floor and correcting the dampness problem.

Perhaps you want to build a small office in your attic, but the cost of running ductwork to the space would be more than the space is worth to you. Again, insulate the area well and add an electric baseboard heater. The installation cost will be low and you can turn the heater off when the room is not occupied.

A weekend cabin which will be used sporadically is another good application for electric heat. By using electric heaters in a cabin, you eliminate the need for fuel tanks and chimneys, which cuts your initial cost to a fraction of that required for other types of heating systems. Since you won't be using the cabin very often during cold weather, the operating cost should not eat you up, but you do have a comfortable heating system when you need it.

Chapter 4 of this book describes in detail how to select and install electric heating units of all kinds.

COMBINATION HEATING AND COOLING UNITS

The advantages of thru-wall heating and cooling units are similar to those of electric baseboard heaters except that they have cooling capability also. Besides all-electric models, units with hot-water heating coils are available. They are especially useful to home owners who want to replace their old steam or hot-water radiators. Chapter 6 gives all the details necessary for installation of both electric and hot-water thru-wall heating and cooling units.

HIGH-VELOCITY HEATING AND COOLING SYSTEMS

High-velocity air systems have been used in commercial buildings for quite some time, but due to the noise once common to this type of system, very few were installed in residential buildings. Now, however, new designs in this type of system have reduced the noise level to the point where it's quite acceptable for residential applications.

The compactness of the equipment and the small ductwork and outlets make this system well-suited for use in existing structures where the installation of conventional forced-air ducts would require too much cutting and patching. On the other hand, the small (3 1/2-inch diameter) ducts of a residential high-velocity system can be fished through wall partitions, corners of closets, and similar places. The system's small air outlets (only 2-inch openings) can be placed nearly anywhere and still provide good air distribution.

The ease of installation of a high-velocity system more than offsets the higher cost of the components; Chapter 9 gives further information on the system should you want to install one.

HEAT PUMPS

A *heat pump* is a system in which refrigeration equipment takes heat from a source and transfers it to a conditioned space (when heating is desired), and removes heat from the space when cooling and dehumidification are desired.

The most common type is the *air-to-air* heat pump, where outside air is used as both a heat source and heat-discharge medium. This type of heat pump can extract heat from the air at almost any temperature—even below zero—and move it inside the conditioned space.

Although heat pumps are more expensive than conventional forced-air heating systems, they soon pay for themselves because they have the unique ability to furnish more heat energy than they consume in electricity. Their ratio of useful heat output to electric input is two or more to one!

Another type of heat pump is the *water-to-air* system. This type of heat pump extracts heat from circulating water and

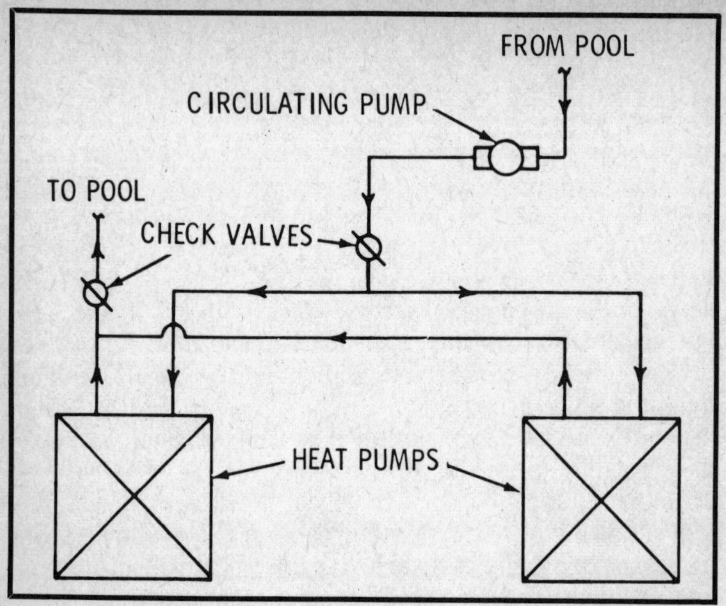

Fig. 3-2. Diagram of piping connecting two water-to-air heat pumps.

transfers the heat to the conditioned space (when heating is desired), and extracts the heat from the conditioned space and transfers it to the circulating water (when cooling is desired).

This type of heat pump was recently used in a large residence which had an indoor swimming pool. In this application, two water-to-air heat pumps were connected as shown in Fig. 3-2. The pool water was used as a heat source and a heat sink. Since this water was preheated to approximately 78°F, the efficiency of the heat pumps approached the maximum.

During the warm months the heat pumps were reversed (cooling cycle) to cool and dehumidify the living area, using the pool water to absorb the heat extracted. This resulted in a system that cooled the living space during warm months and also heated the pool water to a comfortable temperature. The air from the heat pumps was distributed by means of underfloor ducts with supply-air diffusers and return-air grilles mounted in a wainscot.

A similar installation used a surface spring with a constant flow of 52°F ground water as the heat source and sink. In both cases, a noticeable savings in fuel cost was noted.

The installation of ductwork for a heat-pump system is exactly like that of any forced-air system. You can even add an automatic humidifier and an electrostatic air filter if you desire a first-class comfort-conditioning system.

As mentioned previously, the initial cost of a heat-pump system is more than that of a comparable conventional system. Also, you may find that service is a problem. The reason for this is that when heat pumps first arrived on the market, all of the "bugs" had not been taken care of, and many contractors became so fed up with these units that they quit installing and servicing them altogether. Now, however, the bugs have been worked out. Many of these same contractors are getting back into heat pumps, so it should not be too long before good service is available.

SOLAR HEAT

A headline in the December 26, 1959, issue of the 9.e *Washington Star* stated: "Mr. Thomason spends 30 cents for fuel so far this year." The statement was true. He had only used two gallons of fuel oil (at 15 cents a gallon) to supplement the heat produced by the solar system in his home. His fuel cost for the entire year was only $6.30!

Today, thousands of homes are heated with solar energy and, although the initial cost runs two, three, or more times that of a regular heating system, the savings in fuel are out of this world.

A basic solar system consists of a solar (heat) collector that is usually arranged so it faces south; a heat reflector mounted on the ground in front of the collector; a storage tank to hold the sun-heated water; a circulating pump; and piping. The operation of these components is simple. The storage tank is filled with water which is pumped to the top of the heat collector. As the water flows over the collector, the sun heats it. At the bottom of the collector, the heated water is collected in a trough, and then flows back to the storage tank.

The water may be pumped from the storage tank through a system of pipes to baseboard radiators in the living area; a heat exchanger in the water loop may transfer the heat to forced air for distribution via a duct system; or a heat pump may perform the exchange, using the water loop as both heat source and sink.

During warm weather, the water may be pumped over the opposite side of the heat collector—at night—for cooling.

If you think you would like to tackle a solar heat installation, there are many designs and plans available. One source is Edmund Scientific Co., 300 Edscorp Building, Barrington, N.J. 08007.

In most areas of the United States, solar-heated homes require some auxiliary system to provide heat when the weather is persistently cloudy. This auxiliary heat could consist of any system described in this chapter, but usually electric duct heaters are preferred. They are inserted in the ductwork of the system and interlocked with the solar heating controls. When the air temperature goes below a specified point, the duct heaters energize and operate until the solar system again takes over.

Chapter 4

Installing Electric Heating Units

Two decades ago, electric heating units were used only for supplemental heat in small, seldom-used areas of the home, or in weekend or vacation homes. Now, however, electric heat in both new and renovated homes is common.

In addition to the fact that electricity is the cleanest fuel available, electric heat is usually the least expensive to install. Individual room heaters are very inexpensive when compared to furnaces and ductwork; no chimney is required; no utility room is necessary since there is no furnace or boiler; and the installation time and labor are less. Combine all of these features and we have a heating system that is made to order for the do-it-yourselfer.

So, if you are planning the heating system for an addition to your home or, for that matter, your entire home, individual electric heating units may be the answer.

SUPPLEMENTAL ELECTRIC HEATERS

Let's consider a heating and cooling system for a new home using a central unit with ductwork run to the various areas in the home. However, certain areas—like the bathrooms—may be more comfortable if they are not cooled during warm months; yet, they most certainly will have to be heated during the winter months. Dampers may be installed in the bathroom ducts to shut off the air supply during summer months. Another solution is to install a conventional central system with ductwork run to all areas except the bathrooms; there, install radiant electric heaters. During the winter months, no air from the central system enters the bathrooms to cause an uncomfortable draft when you step out of the shower. In summer months, no blast of air conditioning will raise goose bumps while you shave.

Perhaps you are remodeling your attic—turning it into additional bedrooms—but you find that your present forced-air heating system is not large enough to handle the additional load. Rather than install a completely new system, you can merely add a couple of electric baseboard units in the new rooms.

Or maybe you want to replace your present gravity-fed coal heating system with a cleaner type of heat but you do not want to go to the expense of installing a complete duct system throughout the house. Again, electric heating units could be the answer.

PLANNING THE SYSTEM

Regardless of the area you plan to cover with an electric heating system, you can "size" the unit or units by making heat-loss calculations as described in Chapter 2. If you plan to install a supplemental electric heating unit in a room, calculate the heat loss for this room only.

Once the sizes of the units are determined, your next step is to locate the units where they will do the most good and, in the case of existing homes, where wiring will be the easiest to install.

The best type of electric heating equipment for a given area depends on the structural conditions, the purpose for which the room will be used, and the owner's preferences. Electric heating equipment may be mounted along the baseboard; on or in the wall, ceiling, or floor; in kick spaces under counters; and in forced-air duct systems. In addition, electric furnaces which can be mated with air or water distribution systems are available.

The average home owner will probably find baseboard heating units the most practical. They're easy to install and seem to be the most popular for use in homes. Wall-mounted fan-type heating units will probably be second, and ceiling-mounted units will be third. The best way for the home owner to determine which is best is to study manufacturers' catalogs and read the description of each unit. (Manufacturers' names and addresses are given in Appendix I.) It stands to reason that a fan-forced heater would be undesirable where noise may cause a problem. On the other hand, the fan-forced heater may be just right for use in a laundry or bathroom because of its ability to raise room temperature quickly.

For example, suppose a laundry room will be used only twice a week. There is no need to keep the room at 70°F during the long periods when it is not in use. Therefore, the thermostat is turned back to, say, 45°F when the room is not in use, and turned up to 70°F a few minutes before the room will be used on laundry days. A few minutes is all that is required to bring the room up to a comfortable temperature with a fan-forced heater, whereas a radiant baseboard heater would take at least three times as long. From this we can see that the selection of the right type of electric heating unit is very important.

In most cases, electric heating units should be located near where the most heat loss occurs in a room; e.g., under a window on an outside wall. However, structural conditions sometimes prevent locating the units ideally. As mentioned previously, the routing of the wiring is also an important consideration when locating the units, as they certainly won't be of any use if you can't supply them with electricity.

CALCULATING THE CIRCUIT SIZE

Most electric heating units used in residences require 240 volts, although there are many portable heaters designed to operate on 120-volt circuits. Many home owners are somewhat reluctant to tackle the wiring of 240-volt circuits, probably due to fear of being shocked or the misconception that 240-volt circuits are much more complicated than 120-volt ones.

There is no more danger working on a dead 240-volt line than on a dead 120-volt line or even a dead flashlight circuit. Electricity must always be handled with care; carelessness with 120 volts can kill you just as dead as 240 volts. On the other hand, if proper precautions are taken before working on an electrical line, there is little danger involved. Just make certain that the *right* fuse is pulled or the *right* circuit breaker is tripped before working or handling any electrical wiring; then double-check the circuit—at both ends—with a test lamp or instrument before touching any electrical component which could be live.

240-volt circuits and equipment are usually no more complex than those operating at 120 volts. In fact, most connections to 240-volt electric heating equipment involve only two wires, much like a lighting fixture or a duplex receptacle.

With certain restrictions, the wiring methods used for circuits feeding electric heating units will be identical to other branch circuits in your home. These include nonmetallic cable, armored cable, conduit, wiremold, etc. The wires, of course, must be of size sufficient to meet the requirements set forth in the National Electrical Code, and are determined on the basis of the equipment load requirements and the length of the run. The proper overcurrent protection (fuse or circuit breaker) must be provided.

For example, to size the circuit wire to feed an electric baseboard heater installed in a newly remodeled bedroom, proceed as follows:

1. Find the nameplate on the equipment and record the data. Let's assume a rating of 1.2 kW at 240 volts.
2. Calculate the load in amperes using the formula

$$\frac{\text{kW} \times 1000}{\text{volts}} = \text{amps}$$

$$\frac{1.2 \times 1000}{240 \text{ volts}} = \frac{1200 \text{ watts}}{240 \text{ volts}} = 5 \text{ amps}$$

3. Select the proper wire size from the following table. Either copper or aluminum wire may be used.

Size 14 AWG wire (rated at 15 amperes) is the smallest allowable size for use on branch circuits in residential wiring. Even though the heater in the example draws only five amperes of current, size 14 AWG copper (or 12 AWG aluminum wire) is the smallest size permissible. Of course, more than one heater may be fed from one circuit provided that the total load of all heaters does not exceed 80% of the current-carrying capacity of the wire; that is, on 14-gauge wire, the maximum load should not exceed 80% of 15 amperes, or 12 amperes.

Table 4-1. Current-Carrying Capacity of Copper and Aluminum Wire.

Wire Size	Max. Load in Amps	Max. Load in Amps.
	copper wire	aluminum wire
14 AWG	15	NA
12 AWG	20	15
10 AWG	30	25
8 AWG	40	30
6 AWG	55	40
4 AWG	70	55
3 AWG	80	65
2 AWG	95	75
1 AWG	110	85

As mentioned previously, there are several different wiring methods that may be used for circuits feeding electric heating units. However, make certain that an equipment ground is provided for all circuits. If nonmetallic (NM) cable is used, purchase two-wire cable of the proper size with a third *grounding* wire. In the case of armored cable (BX) or conduit, the metal jacket enclosing the wires acts as a path to ground.

In most cases, one end of the circuit will be connected to the proper size 240-volt circuit breaker in the house panelboard. The black and white wires are connected to terminals on the circuit breaker and the bare grounding wire is connected to the solid neutral bus bar in the panel enclosure. From the panel, the cable is run to a thermostat or other means of control, and then to the terminals of the heating unit itself (Fig. 4-1). The exact connection procedures will vary from heater to heater, but the instructions given later in this chapter will enable you to connect any type properly.

To summarize, first determine the size of wire required, run the cable, make connections at the equipment and at the panel or other disconnecting device, run the circuit through the thermostat or other control device, take precautions against electrical shock, and make all terminal connections secure.

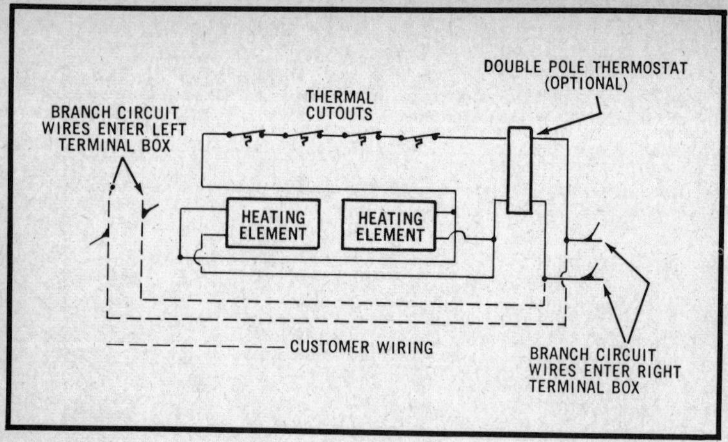

Fig. 4-1. Wiring diagram of a typical electric baseboard heater.

With the basic wiring out of the way, we now look at the various electric heating units available. The description of each will help you decide which will best suit your needs. The installation instructions following each type will show how to install it.

ELECTRIC BASEBOARD HEATERS

A typical electrical baseboard heater is shown in Fig. 4-2. As the name implies, this type of heater is mounted on the floor along the baseboard, preferably on outside walls under windows. It is a quiet, efficient heater and is the type most often used for residential electric heat. These units may be mounted on practically any surface, but if polystyrene foam insulation is used near the unit, a 3/4-inch (minimum) ventilated spacer strip must be used between the heater and the wall. Solid spacers such as wood, plaster, etc., should not be used. In such cases, the heater should also be elevated above the floor or rug to allow ventilation to flow from the floor upward over the total heater space. Also be sure to check the voltage rating of your particular unit to ascertain that it is the same as the electric supply voltage in your home. Finally, all wiring should be in accordance with local codes and the National Electrical Code.

Unpack the heater from its carton and remove the heater cover by first prying loose the lower edge at the end and then pulling outward on the lower edge; store the cover in the carton until it is needed. Some heaters will have screws to

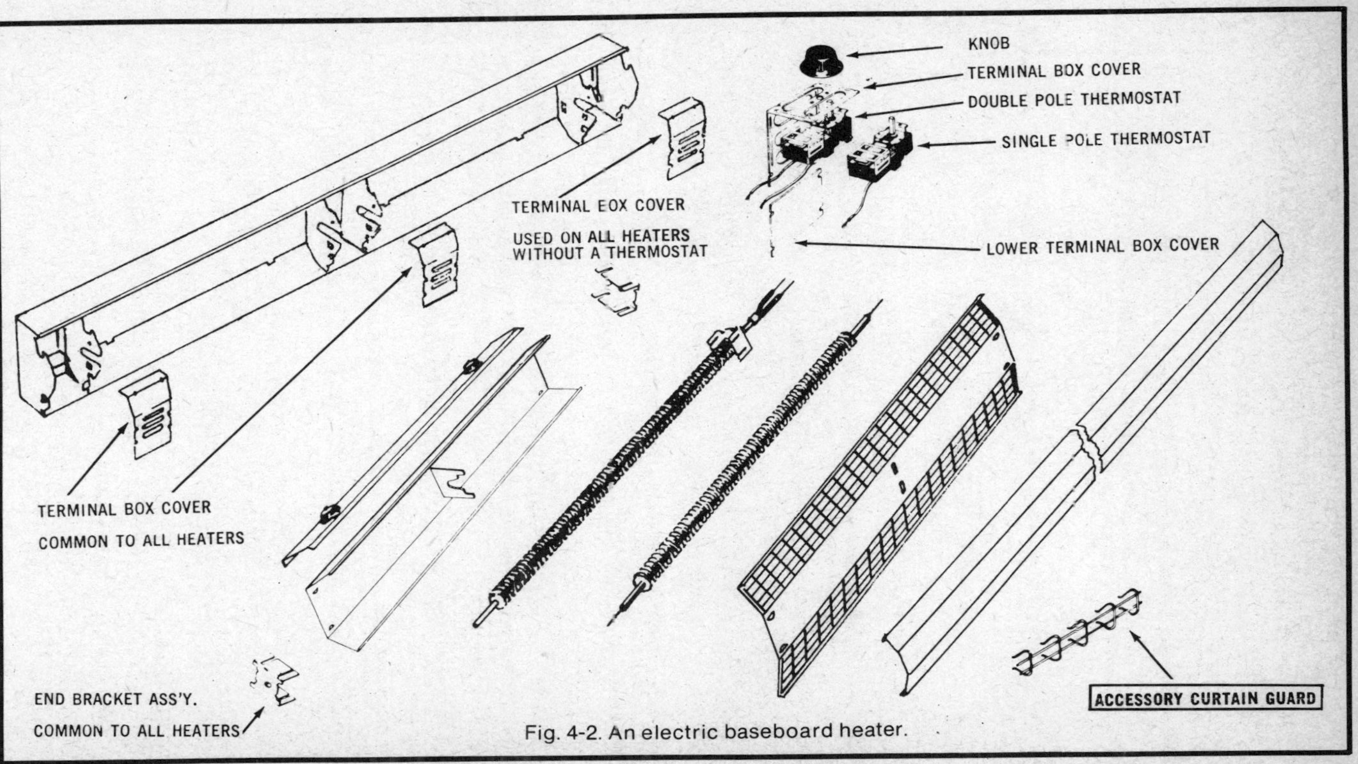

Fig. 4-2. An electric baseboard heater.

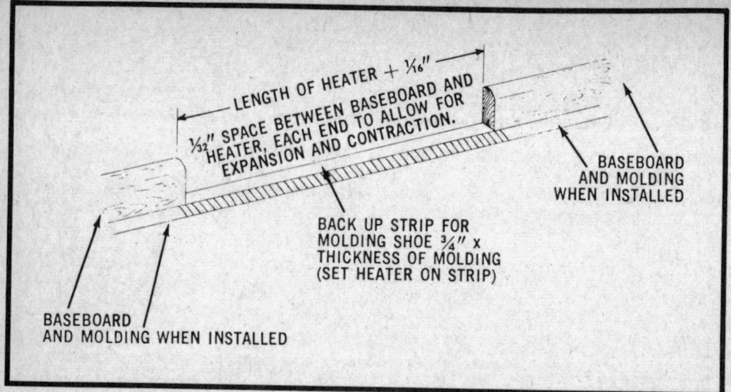

Fig. 4-3. Procedures for preparing a new home for the installation of a baseboard heater when baseboard molding is to be used.

remove before the cover can be removed. Then remove the cover of the terminal box on the end which the wiring will enter.

Position the unit at the desired location on the wall and remove the appropriate knockout from either the bottom or back of the heater to allow the wire and wire connector to be inserted later. Set up the heater as shown in Fig. 4-3 and Fig. 4-4 (for new construction) and Fig. 4-5 (for old construction). Mark heater mounting holes and proper knockout hole.

Drill a half-inch hole at the knockout location and at least two quarter-inch mounting holes; then rough-in electrical wiring to the knockout and secure the wire with the proper wire connector.

Fasten the heater back panel to the wall using one of the methods shown in Fig. 4-6, and then make electrical connections as shown in Fig. 4-1. The thermostat may be either

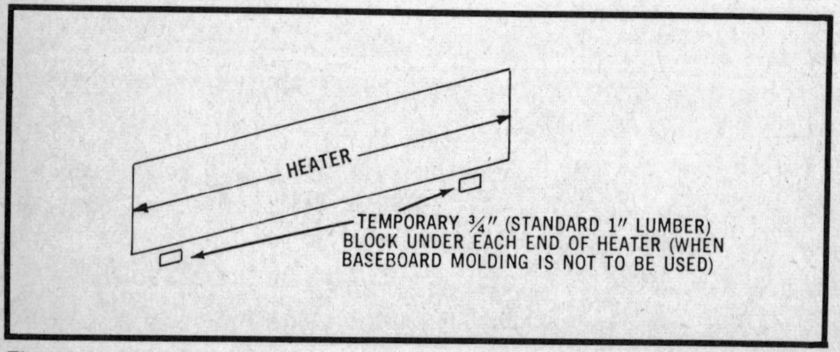

Fig. 4-4. Method used to prepare for the installation of an electric baseboard heater when baseboard molding is not to be used.

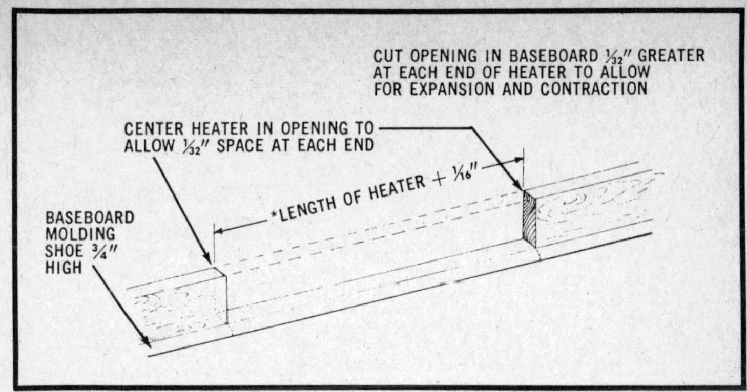

Fig. 4-5. Method of preparing an existing house for the installation of electric baseboard heaters.

wall-mounted or mounted directly on the heater, although the former will give better results.

The procedures given so far pertain mainly to heaters controlled by *line-voltage* thermostats, which operate using 240 volts and must be wired with cable sufficiently heavy to carry all the load drawn by the heater. Not all electric heat units use line-voltage thermostats; some thermostats operate on reduced voltage which is supplied by a transformer, thus enabling you to use lighter wire on the thermostat circuit. This is financially advantageous in situations which require long runs of thermostat wire.

If the thermostat you select is low-voltage, it will come with a transformer and a relay. Follow the instructions packed

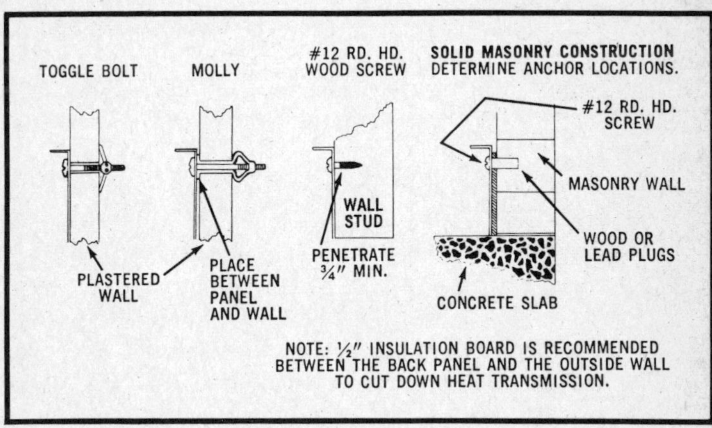

Fig. 4-6. Suggested ways of fastening heater to common wall surfaces.

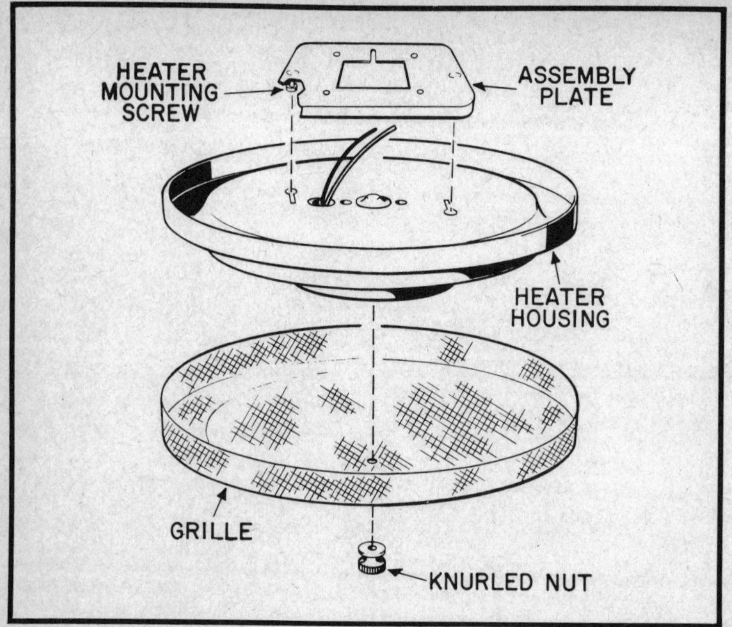

Fig. 4-7. Pictorial drawing of a radiant ceiling heater installation.

with the unit for installation. Never use wire intended for use with a low-voltage thermostat on a line-voltage thermostat. Circuit breakers will not protect you under this circumstance, and the probable result will be a fire.

Replace the terminal box cover and snap the heater cover in place before making the connections at the electrical source or panelboard. When you are certain that all electrical connections are secure and proper, turn on the circuit breaker and turn the thermostat to a setting above room temperature to test the heater. If all wiring and connections are correct, the heater should become warm within five minutes.

RADIANT CEILING HEATERS

Ceiling heaters are often used in bathrooms and similar areas so that the entire room does not have to be overheated to meet the need for extra warmth after a bath or shower. They are also used in larger areas, as a garage or basement, or for spot-warming a person standing at a workbench.

Most of these units are rated from 800 to 1300 watts and normally operate on 120-volt circuits. As with all electric heating units, they may be controlled by a remote thermostat, but since they are usually used for supplemental heat, a conventional wall switch is usually used. They are quickly and easily mounted on an outlet box (see Fig. 4-7) in similar fashion to a lighting fixture. Run a 120-volt circuit from the heating unit, through a control device, and connect to a single-pole overcurrent device in the panel.

RADIANT HEATING PANELS

Radiant heating panels are commonly manufactured in two-by-four-foot sizes and are rated at 500 watts at 120 or 240 volts. They may be located on ceiling or walls to provide radiant heat which spreads evenly throughout the room. Each room may also be controlled by its own thermostat. Since this type of heater may be mounted on the ceiling, their use allows complete freedom of room decor, furniture placement, and drapery arrangement. Most are finished in a snow-white texture to blend with white plaster finishes, or they may be painted to complement any color scheme.

Units mounted on the ceiling will give the best results when located parallel to, and approximately two feet from, the outside wall. Use the cardboard template provided on the carton of each unit to drill the mounting holes (Fig. 4-8) in the ceiling. If necessary, an access hole adequate for mounting the junction box can be cut inside the mounting surface in any convenient location. Just make certain that this cut is within the margin shown on the mounting plate.

Next, remove the template and install the mounting clips in the same orientation as shown on the template. Fasteners provided with each panel may be used through plaster and lath, wallboard, or into wood framing members as shown in Fig. 4-9.

Fig. 4-8. Using the cardboard template to drill mounting holes for radiant heating panels.

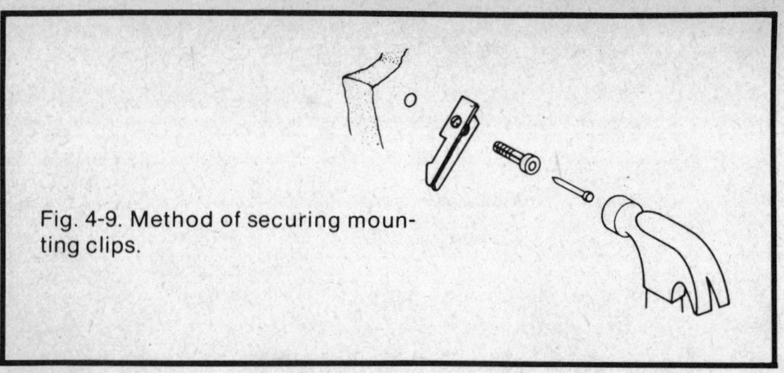
Fig. 4-9. Method of securing mounting clips.

The junction box is securely mounted on the opposite side of any insulation and as far from the panel as possible. For the supply connection, use wire which has insulation rated for at least 90°C and make certain that the junction box is grounded. The box may be mounted with wood screws to any nearby wood framing member (Fig. 4-10) and then connected as described in the applicable wiring diagram in Fig. 4-11.

Raise the panel flush with the mounting surface, ensuring that all four mounting clips insert into their appropriate slots in the heating panel. This is easily done by inserting two clips on one side of the panel and then swinging the panel flush to engage the other two clips. When you are certain that the panel is flush with the mounting surface, slide the panel approximately one inch toward the end where the wire connections is made. This locks the panel in place. See Figs. 4-12 and 4-13.

This type of heating panel may also be installed on a wall using the same methods as described for mounting on the ceiling.

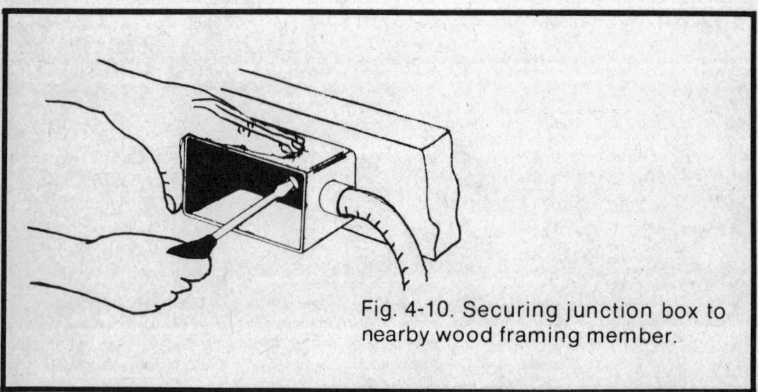
Fig. 4-10. Securing junction box to nearby wood framing member.

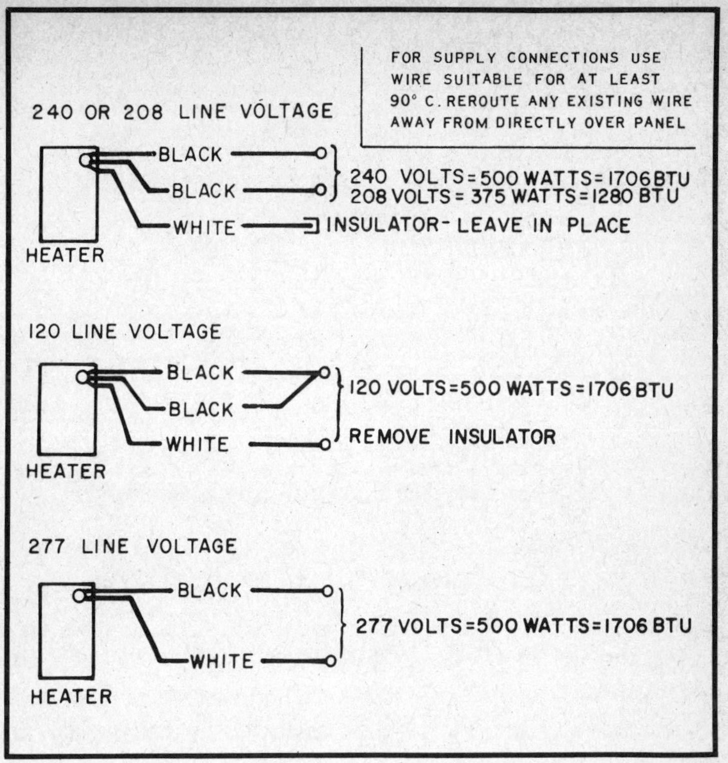

FOR SUPPLY CONNECTIONS USE WIRE SUITABLE FOR AT LEAST 90° C. REROUTE ANY EXISTING WIRE AWAY FROM DIRECTLY OVER PANEL

240 OR 208 LINE VOLTAGE

BLACK
BLACK
WHITE
HEATER
240 VOLTS = 500 WATTS = 1706 BTU
208 VOLTS = 375 WATTS = 1280 BTU
INSULATOR- LEAVE IN PLACE

120 LINE VOLTAGE

BLACK
BLACK
WHITE
HEATER
120 VOLTS = 500 WATTS = 1706 BTU
REMOVE INSULATOR

277 LINE VOLTAGE

BLACK
WHITE
HEATER
277 VOLTS = 500 WATTS = 1706 BTU

Fig. 4-11. Wiring diagram for radiant heating panels of various supply voltages.

ELECTRIC INFRARED HEATERS

Infrared rays are like the rays of the sun. Practically speaking, they do not heat the air through which they travel; rather, they heat only persons and certain objects which they strike. Therefore, infrared heaters are designed to deliver heat into controlled areas for

Fig. 4-12. Raising the panel for mounting.

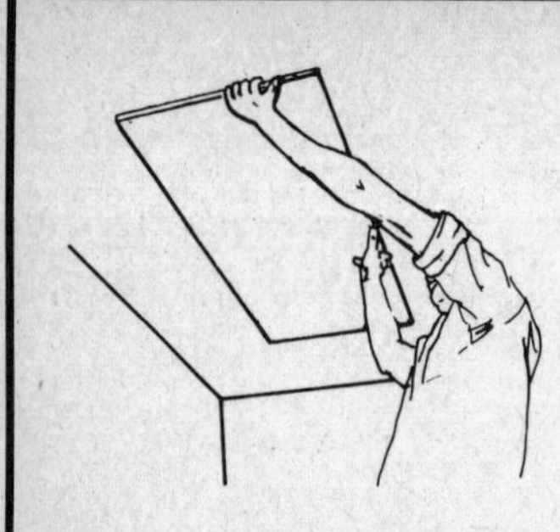

Fig. 4-13. Sliding the panel approximately one inch toward the end where the wire connection is made to lock the panel in place.

the efficient warming of people and surfaces both indoors and outdoors. The advantages of such units include:

1. Heat rays are confined to the desired area—no waste—no loss.
2. No warm-up period is required.
3. They are easy to install as no ducts, vents, or plumbing are required.
4. The infrared quartz lamps provide some light in addition to heat.

Infrared heating units are ideal for heating a person working at a workbench without heating the entire room; for melting snow from porches during winter months; for sun-like heat over indoor or outdoor swimming pools; or for heating people relaxing on patios during cool fall or early spring days.

Most types of infrared fixtures are designed for installation as illustrated in Fig. 4-14. In no case should the heater be mounted closer than 24 inches from vertical walls. For mounting on the surface of a ceiling, proceed as follows:

1. Remove the mounting screw at the end of the wireway that holds the wireway to the canopy.
2. Slide the wireway approximately half an inch toward the opposite end from the screw just removed. Then raise the wireway and remove it completely.
3. Mount the wireway to the ceiling, attaching the electrical connector to the top or ends of the wireways as shown in Fig. 4-14.
4. Pull required wiring into the wireway using wire rated for at least 75°C. Refer to the wiring diagram packed with your particular heater.

5. Attach the heater to the wireway at the end opposite from the wireway attachment screw, and let it hang at a 90° angle from the tab on the end of the wireway as shown in Fig. 4-15.
6. Complete the heater wiring, using wire provided in the wireway, to field or circuit wiring.
7. Rotate the heater up to its horizontal position and slide it half an inch toward the end that contains the mounting screw; then install the mounting screw for a completed job.

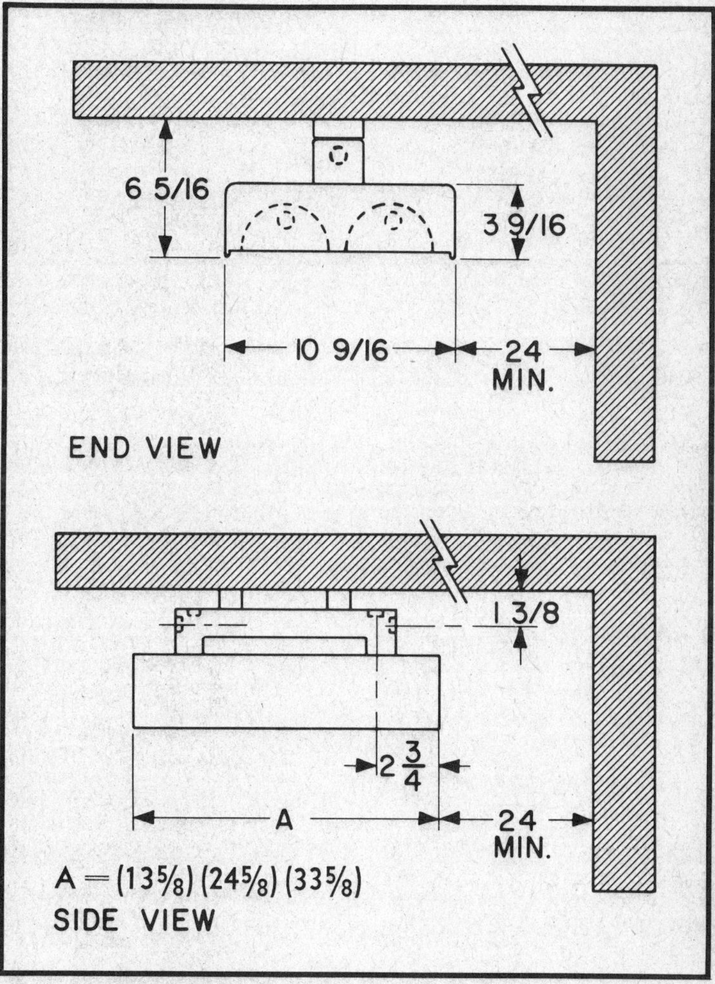

Fig. 4-14. Mounting distances for infrared heaters.

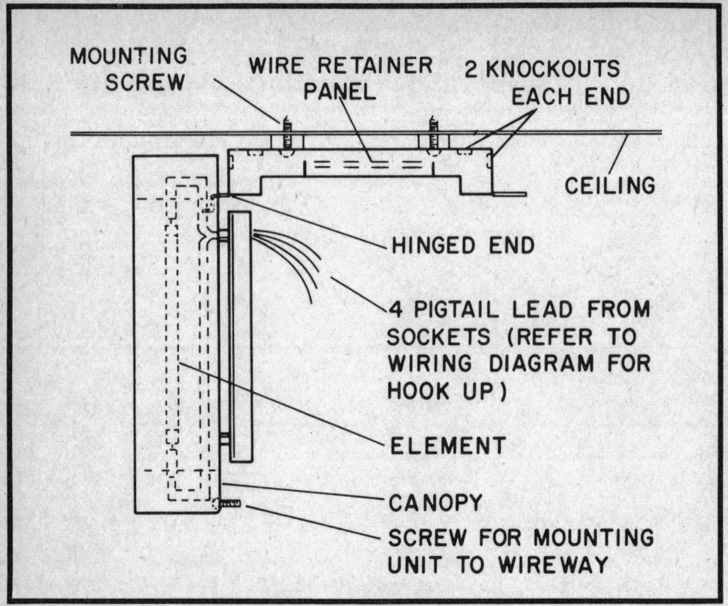

Fig. 4-15. Attaching the heater to the wireway.

Infrared fixtures intended for outdoor use are designed for installation as illustrated in Fig. 4-16 and, as shown in the drawing, should never be mounted closer than 18 inches from vertical walls. A minimum of 5 inches is required between the ceiling and the top of the heater. To prepare for installation:

1. Check the heater nameplate voltage to make sure it is the same as your electrical service.
2. Feed the heater lead wire through the conduit nipple (supplied with the heater) and screw the nipple to the hub on top of the heater.
3. Attach the outlet box cover to the conduit nipple.
4. Connect the electrical circuit wires to the heater leads and secure the cover to the outlet box.

Install the proper heating element—quartz lamp, quartz tube, or metal-sheathed lamp—in the heating unit as recommended by the manufacturer. A heavy-duty single-pole switch is the control normally used for this type of heater.

FORCED-AIR WALL HEATERS

This type of electric heater is designed to bring quick heat into an area where the sound of a quiet fan will not be disturbing—the den, recreation room, or both are examples. Most of these units are equipped with a built-in thermostat with a sensor mounted in the intake air stream. Some types are available for mounting on high walls or even ceilings.

Some of the wall heaters are designed to fit in walls of standard frame construction, with two-by-four studs on 16-inch centers, while others have to be framed in with additional members. Figure 4-17 shows the installation procedures for one type of electric wall heater.

First cut an opening in the wall a minimum of six inches above the finished floor and at least six inches from any

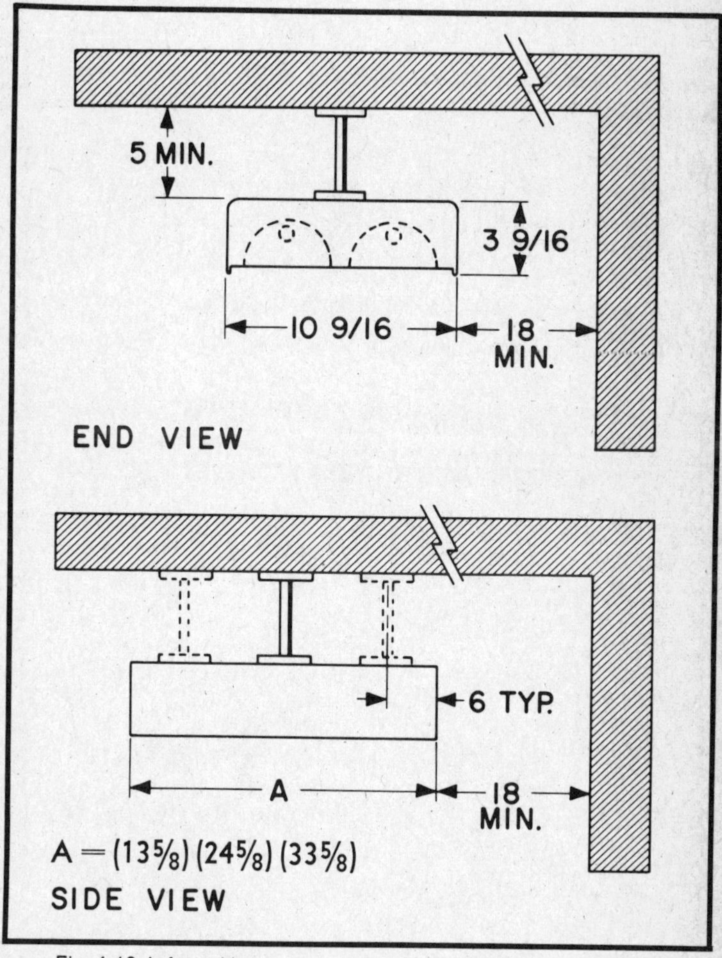

Fig. 4-16. Infrared heater mounting dimensions for outdoor use.

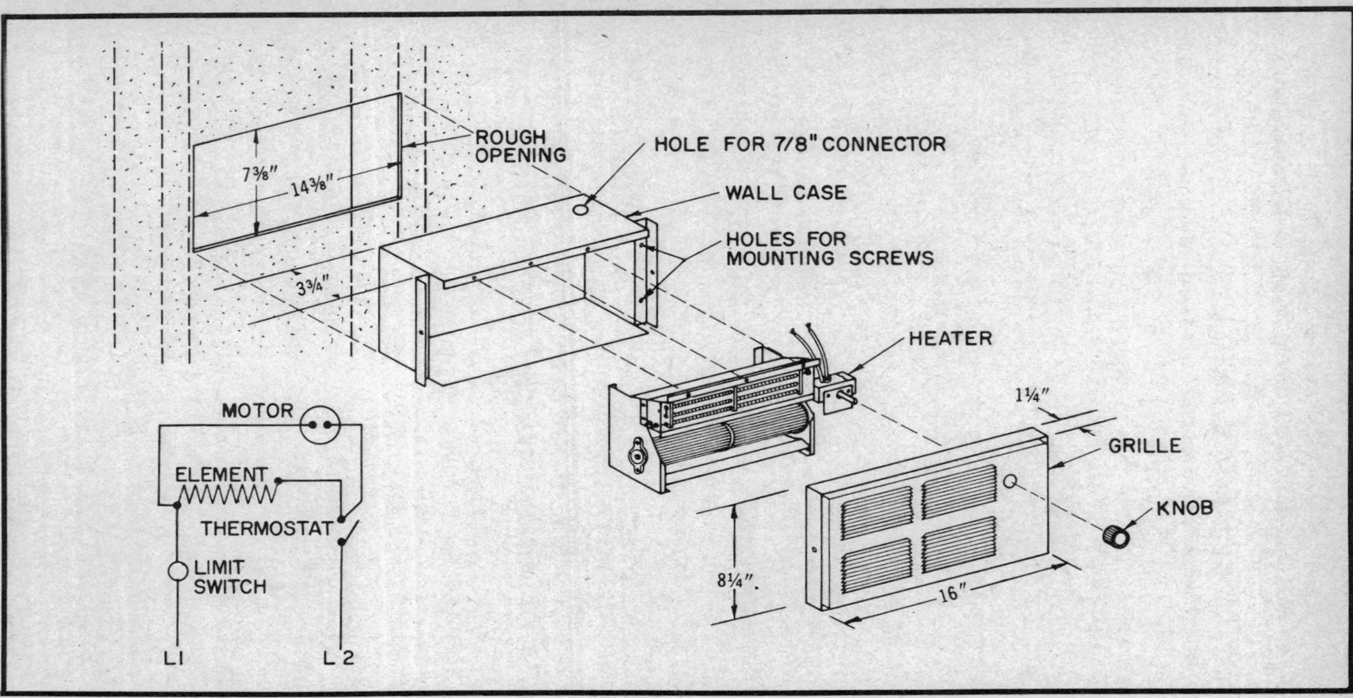

Fig. 4-17 Installation procedures for one type of electric wall heater.

adjacent wall. You will obtain more efficiency from the heater if you mount it as low as possible.

Disassemble the unit as shown in the drawing and, after making sure the current is off, run a two-wire circuit with ground wire from the panelboard to the heater location. Insert the heater end of this circuit wire through the 7/8-inch hole in the case and install a 7/8-inch clamp or wire connector to secure the cable.

Insert the case of the heater into the wall opening, pushing it back until the flanges on the case rest against the wall. Then secure it with nails through the holes provided.

Connect the ground wire to the grounding stud provided on the case of the heater before connecting the remaining circuit wires to the two leads on the inner unit of the heater. Then insert the inner unit into the case and push it back into place. Secure the inner unit to the flange at the top of the case with the screws provided.

Place the cabinet grille over the front of the case and secure it in place. Placing the thermostat knob over the shaft on the cabinet front completes this end of the installation. The only remaining work is to connect the panel-end wires to an overcurrent device and turn the unit on.

Other types of fan or forced-air heating units are available for recessed mounting (as shown), semirecessed mounting, and surface mounting.

FLOOR INSERT CONVECTOR HEATERS

This relatively new type of electric heater requires no wall space and is ideally suited for placement beneath conventional or sliding glass doors to form an effective draft barrier. All are equipped with a safety device (thermal cutout) to automatically disconnect the heating element in the event that normal operating temperatures are exceeded (as when a piece of furniture, a rug, or similar article covers the heater accidentally). The heating element itself is suspended in the metal housing between baffles to assure proper balance between downward-flowing return air and upward-flowing heated air.

Floor insert convector heaters may be installed in both old and new houses. Merely cut through the flooring, insert the metal housing and then nail the housing through the mounting holes. Once the wiring is installed, the heavy-gauge floor grille

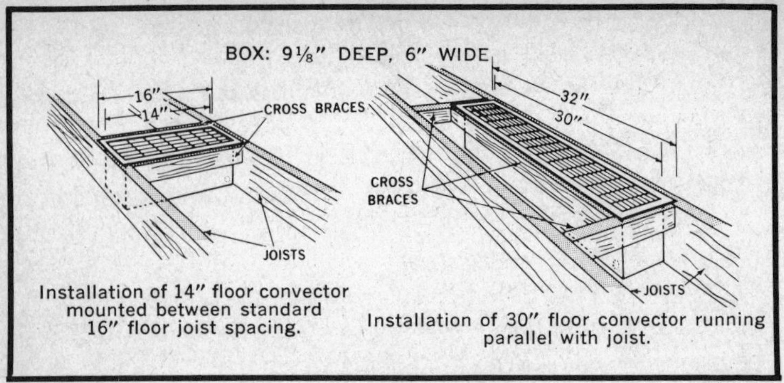

Fig. 4-18. Floor heaters mounted between floor joists and running parallel with the joists.

drops in flush with the finished floor, and is easily removed at any time for cleaning.

Figure 4-18 shows how two different sizes of floor insert heaters are mounted between floor joists. The exact mounting instructions come packed with each piece of equipment.

ELECTRIC KICK-SPACE HEATERS

Modern kitchens contain so many appliances and so much cabinet space for the convenience of the housewife that many times there is no room to install electric heaters except on or in the ceiling. Recently, however, a kick-space heater has been added to the line of versatile electric heating units for inserting into the kick-space area of kitchen counters.

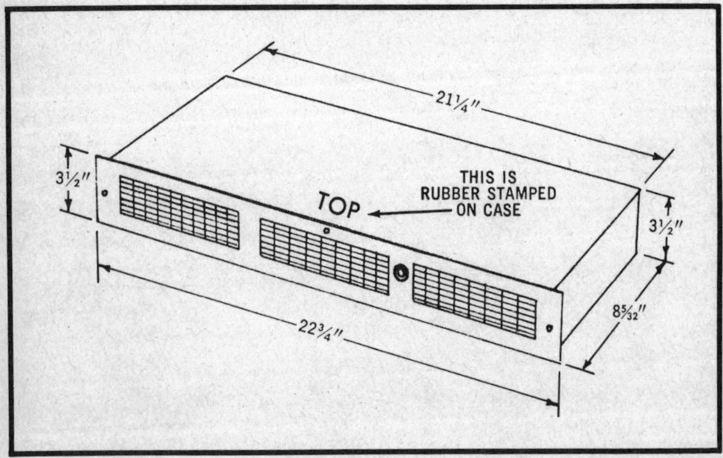

Fig. 4-19. Dimensions of one model of kick-space heater.

For the most comfort, kick-space heaters should not be installed in a manner such that warm air blows directly on occupants' feet. Ideally, the air discharge should be directed along the outside wall adjacent to normal working areas, not directly under the sink.

To begin the installation of this type of heater, remove the grille, fan, and heater assembly from the recess box, and then remove the cardboard packing and tape from the fan and the heater assembly.

Using the back of the carton or packing box as a template, mark the area on the kick space where the unit is to be mounted; the dimensions of one model are shown in Fig. 4-19. Cut the opening in the kick-space and install the foam vinyl seal as shown in Fig. 4-20 before inserting the recess box into the opening (Fig. 4-21).

Some of these units are available with built-in thermostats; others require a wall-mounted thermostat or double-pole switch for control purposes. Run the circuit wires as for any other type of electric heating equipment, sizing the conductors from the nameplate rating and the table presented earlier in this chapter.

After the wiring is roughed-in to the junction box on the heater, connect the power-supply wires from the switch or wall thermostat to the lead wires in the junction box; splice the wires together with wire nuts or other approved connectors. Now partially insert the fan and heater assembly and connect their leads to the power supply. Fasten heater assembly and grille to recess box and you're ready for operation—once the other end of the circuit is connected to your panelboard with the proper overcurrent protection.

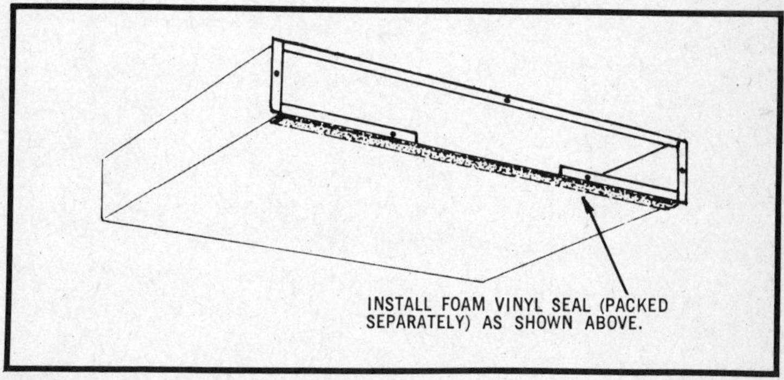

Fig. 4-20. Recess box for kick-space heater with foam vinyl seal installed.

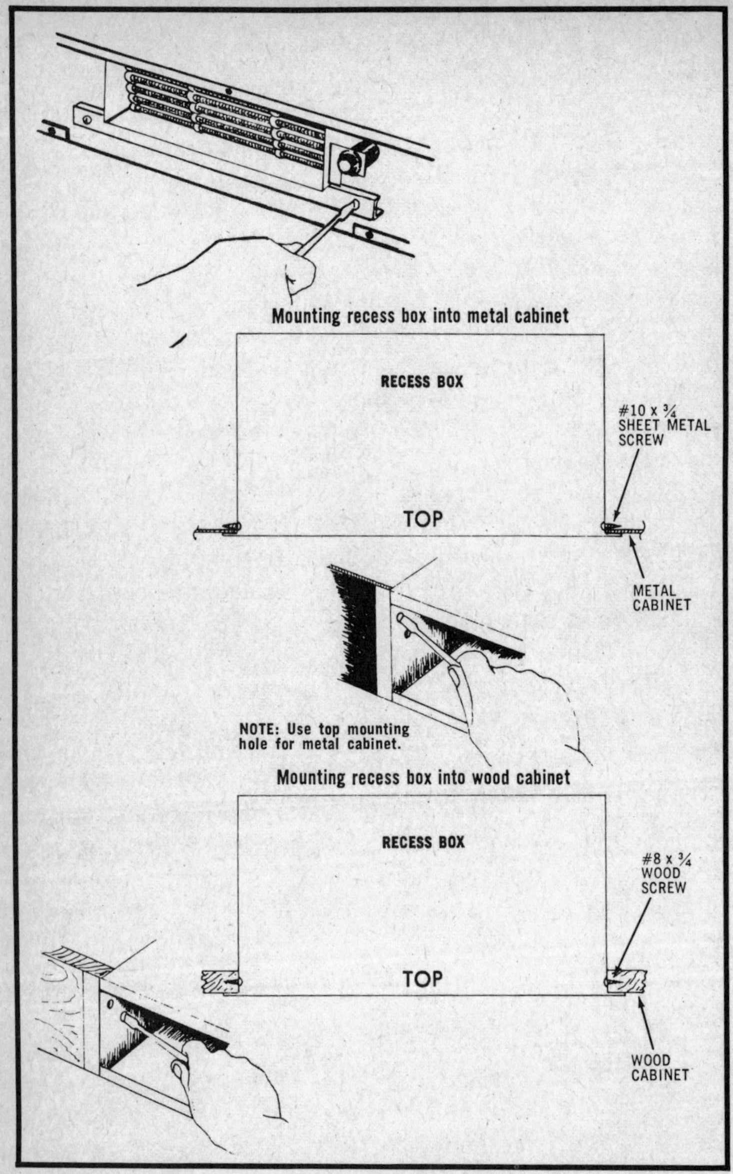

Fig. 4-21. Methods of installing kick-space heater.

About the only maintenance required on this type of heating unit is periodic cleaning of the air filter. A weekly cleaning with a vacuum cleaner will insure a clear passageway for air.

Chapter 5
Simplified Cooling Calculations

An accurate calculation of how much heat enters your home on a hot summer day is essential for proper air conditioning. This calculation dictates the size of components required to cool your home. Care and accuracy at this point pays off in savings and comfort later, so pay careful attention to the instructions given in this chapter and always check your math for errors.

The "Residential Cooling Load Estimate Form" (Fig. 5-1) will be used as a guide to ascertain that all pertinent data will be included.

The two lines at the top of the form should be filled in first; the top line for identification, and the second to provide a basis for accurate calculation in your geographic location. Refer to Table 5-1 to obtain the outside design conditions required to complete the second line of the form. For example, if you live near Washington, D.C., Table 5-1 gives the summer outside temperature as 90° (DB), a *medium* (M) daily range, and a latitude of 40°. Once this data is entered on your form, you are ready to begin the calculations.

We will use the floor plan of the house in Fig. 5-2 as an example. Refer again to the "Residential Cooling Load Estimate Form" in Fig. 5-1.

WINDOWS

Item 1a on the form deals with the amount of solar heat that enters the home through glass surfaces. Begin by

Customer .... Date .... Address ....

Design Conditions .... Outside temperature .... Daily Range .... Latitude ....

1a Windows—Solar Heat (See back side for example)	Area Sq. ft.	No Shades	Roller Shades	Drapes or Venetian	Ventilated Awning	Outside Shade Screens	
(1) Regular Single glass							
(a) North (or Shaded)	......	19	14	11	16	4	......
(b) Northeast and Northwest	......	51	37	28	17	12	......
(c) East and West	......	77	57	43	18	18	......
(d) Southeast and Southwest	......	66	48	36	17	9	......
(e) South	......	36	26	20	16	5	......
(2) Regular Dougle glass							
(a) North (or shaded)	......	17	13	10	11	3	......
(b) Northeast and Northwest	......	45	35	26	12	8	......
(c) East and West	......	67	53	40	12	12	......
(d) Southeast and Southwest	......	57	45	34	12	6	......
(e) South	......	31	25	18	11	3	......
(3) Heat Absorbing Double glass							
(a) North (or shaded)	......	10	8	7	8	2	......
(b) Northeast and Northwest	......	26	22	18	9	7	......
(c) East and West	......	40	33	28	10	10	......
(d) Southeast and Southwest	......	34	28	23	9	5	......
(e) South	......	20	16	13	9	3	......

1b Windows—air to air heat gain	Area Sq. ft	Outdoor Design Temperature 90	95	100	105	110	
(1) all single glass	......	8	12	16	20	24	......
(2) all double glass	......	4	7	9	11	13	......

2 Walls and Doors	Area Sq. ft	Daily Temperature Range: Low		Medium				High				
		Outdoor Design Temperature										
a Frame and Veneer on frame		90	95	90	95	100	105	95	100	105	110	
(1) No insulation	......	5.9	7.2	4.8	6.1	7.4	8.7	4.8	6.1	7.4	8.7	......
(2) less than 1 in. insulation	......	4.3	5.2	3.5	4.5	5.4	6.4	3.5	4.5	5.4	6.4	......
(3) 1 to 2 in. insulation	......	2.9	3.6	2.4	3.1	3.7	4.4	2.4	3.1	3.7	4.4	......
(4) more than 2 in. insulation	......	1.8	2.2	1.5	1.9	2.3	2.7	1.5	1.9	2.3	2.7	......
b Masonry walls 8 in. block or brick												
(1) plastered or plain	......	7.3	9.7	5.4	7.8	10.2	12.6	5.4	7.8	10.2	12.6	......
(2) Furred, no insulation	......	4.6	6.1	3.4	4.9	6.4	7.9	3.4	4.9	6.4	7.9	......
(3) Furred, less than 1 in. insulation	......	3.1	4.1	2.3	3.3	4.3	5.3	2.3	3.3	4.3	5.3	......
(4) Furred, 1 to 2 in. insulation	......	2.1	2.8	1.6	2.3	3.0	3.7	1.6	2.3	3.0	3.7	......
(5) Furred, more than 2 in. insulation	......	1.4	1.8	1.0	1.5	1.9	2.4	1.0	1.5	1.9	2.4	......
c Partitions												
(1) Frame, finished one side only, no insulation	......	8.4	11.4	6.0	9.0	12.0	15.0	6.0	9.0	12.0	15.0	......
(2) Frame, finished both sides, no insulation	......	4.8	6.5	3.4	5.1	6.8	8.5	3.4	5.1	6.8	8.5	......
(3) Frame, finished both sides, more than 1 in insulation	......	2.0	2.7	1.4	2.1	2.8	3.5	1.4	2.1	2.8	3.5	......
(4) Masonry, finished one side, no insulation	......	2.6	4.4	1.2	3.0	4.7	6.5	1.2	3.0	4.7	6.5	......
d Wood doors*	......	11.3	13.8	9.3	11.8	14.3	16.8	9.3	11.8	14.3	16.8	......

Item												
3 Ceilings and Roofs												
a Ceiling under naturally vented attic or vented flat roof												
(1) Uninsulated	dark	......	9.9	11.0	9.0	10.1	11.3	12.4	9.0	10.1	11.3	12.4	......
	light	......	8.1	9.2	7.1	8.3	9.4	10.6	7.1	8.3	9.4	10.6	......
(2) Less than 2 in. insulation	dark	......	4.3	4.8	3.9	4.4	4.9	5.4	3.9	4.4	4.9	5.4	......
	light	......	3.5	4.0	3.1	3.6	4.1	4.6	3.1	3.6	4.1	4.6	......
(3) 2 in. to 4 in. insulation	dark	......	2.6	2.9	2.3	2.6	2.9	3.2	2.3	2.6	2.9	3.2	......
	light	......	2.1	2.4	1.9	2.2	2.5	2.8	1.9	2.2	2.5	2.8	......
(4) More than 4 in. insulation	dark	......	1.7	1.9	1.6	1.8	2.0	2.2	1.6	1.8	2.0	2.2	......
	light	......	1.4	1.6	1.2	1.4	1.6	1.8	1.2	1.4	1.6	1.8	......
b Built up roof no ceiling													
(1) Uninsulated	dark	......	17.2	19.2	15.6	17.6	19.6	21.6	15.6	17.6	19.6	21.6	......
	light	......	14.0	16.0	12.4	14.4	16.4	18.4	12.4	14.4	16.4	18.4	......
(2) 2 in. roof insulation	dark	......	8.6	9.6	7.8	8.8	9.8	10.8	7.8	8.8	9.8	10.8	......
	light	......	7.0	8.0	6.2	7.2	8.2	9.2	6.2	7.2	8.2	9.2	......
(3) 3 in. roof insulation	dark	......	6.0	6.7	5.5	6.2	6.9	7.6	5.5	6.2	6.9	7.6	......
	light	......	4.9	5.6	4.3	5.0	5.7	6.4	4.3	5.0	5.7	6.4	......
c Ceilings under unconditioned rooms		......	2.7	3.6	1.9	2.9	3.8	4.8	1.9	2.9	3.8	4.8	......
4 Floors													
a Over unconditioned rooms		......	3.4	4.6	2.4	3.6	4.8	6.0	2.4	3.6	4.8	6.0	......
b Over basements, enclosed crawl space, or slab on ground		......	0	0	0	0	0	0	0	0	0	0	......
c Over open crawl space		......	4.8	6.5	3.4	5.1	6.8	8.5	3.4	5.1	6.8	8.5	......
5 Outside Air													
a Infiltration, Btuh per sq ft of gross exposed wall area		......	1.1	1.5	1.1	1.5	1.9	2.2	1.6	1.9	2.2	2.6	......
b Mech. ventilation Btuh per cfm		......	16.2	21.6	16.2	21.6	27.0	32.4	21.6	27.0	32.4	37.8	......
6 People (never less than 3)			Number of occupants x 300 Btuh										......
7 Kitchen appliance allowances			1200 Btuh										......
SUB TOTAL			Items 1 thru 7										......
8 Heat gain to ducts			Subtotal Item 1 thru 7 x factor**										......
9 Total Estimated Sensible Heat Gain			Total (Items 1 thru 8)										......
10 Latent heat gain allowance			Item 9 x 0.3										......
11 Total Estimated Design Heat Gain			Item 9 + Item 10 =										......

*Consider glass area of doors as windows. **Factor for ducts in attic with 2 in. insulation .10 Factor for ducts in attic with 1 in. insulation .15; For ducts in slab floors, insulated ducts in enclosed crawl space, unconditioned basement or furred spaces use factor of .05.

Fig. 5-1. Residential Cooling Load Estimate Form.

Table 5-1. Outside Design Conditions for the United States and Canada.

State & City	Winter DB	Summer DB	Daily Range	Summer WB	Latitude Deg.
ALABAMA					
Anniston	10	95	M	78	35
Birmingham	10	95	M	78	35
Gadsden	10	95	M	78	35
Mobile	20	90	L	80	30
Montgomery	20	95	M	78	30
Tuscaloosa	10	95	M	78	35
ALASKA					
Anchorage	−24	—	M	—	60
Barrow	−48	—	—	—	70
Bethel	−43	—	—	—	60
Cordova	−13	—	—	—	60
Fairbanks	−57	—	M	—	65
Juneau	−5	—	—	—	60
Ketchikan	4	—	—	—	55
Kodiak	4	—	—	—	55
Kotzebue	−46	—	—	—	65
Nome	−36	—	—	—	60
Seward	−4	—	—	—	60
Sitka	2	—	—	—	60
ARIZONA					
Bisbee	30	100	H	72	30
Flagstaff	−5	85	H	61	35
Globe	30	105	H	76	35
Nogales	30	105	H	72	30
Phoenix	35	105	H	76	35
Tucson	30	100	H	72	30
Winslow	−5	95	H	65	35
Yuma	40	110	H	78	35
ARKANSAS					
Bentonville	0	95	M	76	35
Fort Smith	5	95	M	76	35
Hot Springs	10	95	M	78	35
Little Rock	10	95	M	78	35
Pine Bluff	10	95	M	78	35
Texarkana	10	100	M	78	35
CALIFORNIA					
Bakersfield	30	105	H	70	35
El Centro	35	110	H	78	35
Eureka	30	90	M	65	40
Fresno	30	105	H	74	35
Long Beach	35	90	M	70	35
Los Angeles	40	90	L	70	35
Montague	15	95	M	70	40
Needles	25	115	H	—	35
Oakland	30	80	M	65	40

State & City	Winter DB	Summer DB	Daily Range	Summer WB	Latitude Deg.
IDAHO					
Boise	−10	95	H	65	45
Idaho Falls	−15	90	H	65	45
Lewiston	−10	95	H	65	45
Pocatello	−15	90	H	65	45
Twin Falls	−15	95	H	65	40
ILLINOIS					
Aurora	−10	95	M	75	40
Bloomington	−10	95	M	76	40
Cairo	0	100	M	78	35
Champaign	−10	95	M	77	40
Chicago	−10	95	M	75	40
Danville	−10	95	M	77	40
Decatur	−10	95	M	77	40
Elgin	−15	95	M	78	40
Joliet	−10	95	M	76	40
Moline	−10	95	M	76	40
Peoria	−15	95	M	76	40
Rockford	−15	95	M	78	40
Rock Island	−10	95	M	76	40
Springfield	−10	95	M	77	40
Urbana	−10	95	M	77	40
INDIANA					
Elkhart	−10	95	M	75	40
Evansville	−5	95	M	78	40
Fort Wayne	−5	95	M	75	40
Indianapolis	−10	95	M	76	40
Lafayette	−10	95	M	76	40
South Bend	−10	95	M	75	40
Terre Haute	−5	95	M	78	40
IOWA					
Burlington	−10	95	M	78	40
Cedar Rapids	−15	95	M	78	40
Charles City	−20	95	M	75	45
Clinton	−15	95	M	78	40
Council Bluffs	−15	100	M	78	40
Davenport	−10	95	M	78	40
Des Moines	−15	95	M	78	40
Dubuque	−15	95	M	78	40
Fort Dodge	−15	95	M	78	40
Keokuk	−15	95	M	78	40
Marshalltown	−15	95	M	78	40
Sioux City	−15	95	M	78	40
Waterloo	−15	95	M	78	40
KANSAS					
Atchison	−10	100	M	76	40
Concordia	−10	95	M	78	40

State & City	Winter DB	Summer DB	Daily Range	Summer WB	Latitude Deg.
MICHIGAN					
Alpena	−10	90	M	75	45
Ann Arbor	−5	90	M	75	40
Big Rapids	−5	90	M	75	45
Cadillac	−10	90	M	75	45
Calumet	−20	80	M	73	45
Detroit	−5	90	M	75	40
Escanaba	−20	85	M	74	45
Flint	−10	90	M	75	45
Grand Haven	−5	90	M	75	45
Grand Rapids	−5	90	M	74	45
Houghton	−20	80	M	73	45
Kalamazoo	−5	90	M	75	40
Lansing	−10	90	M	75	45
Ludington	−5	90	M	75	45
Marquette	−15	80	M	73	45
Muskegon	−5	90	M	74	45
Port Huron	−10	90	M	75	45
Saginaw	−10	90	M	75	45
Sault Ste. Marie	−20	80	M	71	45
MINNESOTA					
Alexandria	−25	85	M	74	45
Duluth	−25	80	M	71	45
Minneapolis	−25	90	M	76	45
Moorhead	−30	95	M	75	45
St. Cloud	−25	90	M	76	45
St. Paul	−25	90	M	75	45
MISSISSIPPI					
Biloxi	25	90	L	80	30
Columbus	10	95	M	78	35
Corinth	5	95	M	78	35
Hattiesburg	20	95	M	80	30
Jackson	15	95	M	78	30
Meridian	15	95	M	79	30
Natchez	15	95	L	78	30
Vicksburg	15	95	L	78	30
MISSOURI					
Columbia	−10	100	M	78	40
Hannibal	−10	95	M	77	40
Kansas City	−10	100	M	76	40
Kirksville	−10	95	M	78	40
St. Joseph	−10	100	M	76	40
St. Louis	−5	95	M	78	40
Springfield	−5	100	M	77	40

Pasadena	40	95	M	70	35
Red Bluff	15	100	H	70	40
Sacramento	30	100	H	72	40
San Bernardino	30	105	H	72	35
San Diego	45	80	L	68	35
San Francisco	35	80	M	65	40
San Jose	40	90	M	70	35
COLORADO					
Boulder	–15	95	M	64	40
Colorado Springs	–10	95	H	65	40
Denver	–10	95	H	64	40
Durango	–5	95	H	65	35
Fort Collins	–15	95	M	65	40
Grand Junction	–5	95	H	65	40
Leadville	–10	95	M	64	40
Pueblo	–15	95	H	65	40
CONNECTICUT					
Bridgeport	0	85	L	75	40
Hartford	0	90	M	75	40
New Haven	0	85	M	75	40
New London	5	85	L	75	40
Norwalk	0	85	L	75	40
Torrington	0	90	M	75	40
Waterbury	0	90	L	75	40
DELAWARE					
Dover	10	90	M	78	40
Milford	10	90	M	78	40
Wilmington	5	90	M	78	40
DIST. OF COLUMBIA					
Washington	10	90	M	78	40
FLORIDA					
Apalachicola	25	95	L	80	30
Fort Myers	40	95	M	78	25
Gainesville	30	95	M	78	30
Jacksonville	30	95	M	78	30
Key West	55	100	L	78	25
Miami	45	90	L	79	25
Orlando	35	90	M	78	30
Pensacola	25	95	L	78	30
Tallahassee	25	95	M	78	30
Tampa	35	95	M	78	30
GEORGIA					
Athens	10	95	M	76	35
Atlanta	10	95	M	76	35
Augusta	20	100	M	76	35
Brunswick	25	95	L	78	30
Columbus	20	100	M	76	35
Macon	20	95	M	78	35
Rome	10	95	M	76	35
Savannah	25	95	M	78	30
Way Cross	25	95	M	78	30
Dodge City	–10	95	H	78	40
Iola	–5	100	M	75	40
Leavenworth	–10	100	M	76	40
Salina	–10	100	M	78	40
Topeka	–10	100	M	78	40
Wichita	–5	100	M	75	40
KENTUCKY					
Bowling Green	0	95	M	78	35
Frankfort	0	95	M	78	40
Hopkinsville	0	95	M	78	35
Lexington	0	95	M	78	40
Louisville	0	95	M	78	40
Owensboro	0	95	M	78	40
Shelbyville	0	95	M	78	40
LOUISIANA					
Alexandria	20	95	L	78	30
Baton Rouge	20	95	M	80	30
New Orleans	25	95	L	80	30
Shreveport	15	95	M	78	35
MAINE					
Augusta	–15	85	L	73	45
Bangor	–20	85	L	73	45
Bar Harbor	–10	85	L	73	45
Belfast	–10	85	L	73	45
Eastport	–10	85	L	70	45
Lewiston	–10	85	L	73	45
Millinocket	–15	85	M	73	45
Orono	–20	85	M	70	45
Portland	–10	85	M	73	45
Presque Isle	–20	85	L	73	45
Rumford	–15	85	L	73	45
MARYLAND					
Annapolis	10	90	M	78	40
Baltimore	10	90	M	78	40
Cambridge	10	90	L	78	40
Cumberland	0	90	M	75	40
Frederick	5	90	M	78	40
Frostburg	–5	90	M	75	40
Salisbury	10	90	M	78	40
MASSACHUSETTS					
Amherst	–5	90	M	75	40
Boston	0	85	M	74	40
Fall River	0	85	L	75	40
Fitchburg	–5	90	M	75	45
Framingham	–5	85	L	75	40
Lawrence	–5	85	L	74	40
Lowell	–5	85	L	74	45
Nantucket	0	85	L	75	40
New Bedford	0	85	L	75	40
Pittsfield	–10	90	M	75	40
Plymouth	0	85	L	75	40
Springfield	–5	90	M	75	40
Worcester	–5	90	M	75	40
MONTANA					
Anaconda	–30	85	H	59	45
Billings	–30	90	H	66	45
Butte	–30	85	H	59	45
Great Falls	–40	90	H	63	50
Havre	–40	95	M	70	50
Helena	–40	90	H	63	45
Kalispell	–30	90	H	63	50
Miles City	–35	95	H	69	45
Missoula	–30	90	H	63	45
NEBRASKA					
Grand Island	–15	100	H	75	40
Hastings	–15	100	M	75	40
Lincoln	–15	95	M	78	40
Norfolk	–15	95	M	78	40
North Platte	–15	100	H	73	40
Omaha	–15	100	M	78	40
Valentine	–20	95	M	78	45
York	–15	95	M	78	40
NEVADA					
Elko	–10	95	H	63	40
Las Vegas	10	110	H	71	35
Reno	5	95	H	65	40
Tonopah	5	90	M	63	40
Winnemucca	–10	95	H	65	40
NEW HAMPSHIRE					
Berlin	–15	85	H	73	45
Claremont	–15	85	M	73	45
Concord	–10	85	H	73	45
Franklin	–15	85	M	73	45
Hanover	–15	85	M	73	45
Keene	–10	85	L	73	45
Manchester	–10	85	M	74	45
Nashua	–10	85	L	74	45
Portsmouth	–5	85	L	74	45
NEW JERSEY					
Asbury Park	5	90	L	78	40
Atlantic City	10	90	L	78	40
Bayonne	0	90	L	75	40
Belvidere	0	90	M	75	40
Bloomfield	0	90	L	75	40
Bridgeton	5	90	L	78	40
Camden	5	90	L	78	40
East Orange	0	90	L	75	40
Elizabeth	0	90	L	75	40
Jersey City	0	90	L	75	40
Newark	0	90	M	76	40
New Brunswick	5	90	L	75	40
Paterson	0	90	L	75	40
Phillipsburg	0	90	M	75	40
Trenton	0	90	L	78	40

Table Continued on Next Page.

Table 5-1. Continued.

State & City	Winter DB	Summer DB	Daily Range	Summer WB	Latitude Deg.
NEW MEXICO					
Albuquerque	10	95	M	65	35
El Morro	0	85	H	65	35
Raton	−5	95	H	65	35
Roswell	5	100	H	71	35
Santa Fe	5	90	M	65	35
Tucumcari	5	95	H	70	35
NEW YORK					
Albany	−10	90	M	74	45
Auburn	−10	90	M	74	45
Binghamton	−5	90	M	72	40
Buffalo	−5	85	M	73	45
Canton	−20	85	M	73	45
Cortland	−10	90	M	74	45
Elmira	−5	90	M	73	40
Glens Falls	−15	90	M	73	45
Ithaca	−5	90	M	73	40
Jamestown	−5	90	M	74	40
Lake Placid	−15	90	M	73	45
New York	5	90	M	76	40
Niagara Falls	−5	85	M	73	45
Ogdensburg	−20	85	M	73	45
Oneonta	−10	90	M	73	45
Oswego	−5	90	M	74	45
Port Jervis	0	90	L	75	40
Rochester	−5	90	M	74	45
Schenectady	−10	90	M	74	45
Syracuse	−10	90	M	74	45
Watertown	−15	85	M	73	45
NORTH CAROLINA					
Asheville	5	90	M	75	35
Charlotte	15	95	M	78	35
Greensboro	10	90	M	76	35
Hatteras	20	90	L	80	35
New Bern	20	95	L	78	35
Raleigh	15	95	M	78	35
Salisbury	10	90	M	78	35
Wilmington	20	90	M	81	35
Winston-Salem	10	90	M	76	35
NORTH DAKOTA					
Bismarck	−30	95	H	73	45
Devils Lake	−30	90	M	70	50
Dickinson	−30	95	H	70	45
Fargo	−30	95	H	75	45
Grand Forks	−30	90	M	72	50
Jamestown	−30	95	M	73	45
Minot	−35	90	M	71	50
Pembina	−35	90	M	73	50
Williston	−35	90	M	73	50
OREGON					
Arlington	5	95	M	68	45
Baker	−15	90	M	66	45
Eugene	15	90	H	68	45
Medford	20	95	H	68	40
Pendleton	−10	90	H	66	45
Portland	10	85	M	68	45
Roseburg	20	90	H	66	45
Salem	15	90	H	68	45
Wamic	0	90	H	66	45
PENNSYLVANIA					
Altoona	−5	90	M	75	40
Bethlehem	0	90	M	75	40
Coatesville	5	90	M	75	40
Erie	−5	85	M	74	40
Harrisburg	5	90	M	75	40
New Castle	−5	90	M	75	40
Oil City	−5	90	M	75	40
Philadelphia	5	90	M	78	40
Pittsburgh	−5	90	M	75	40
Reading	5	90	M	75	40
Scranton	0	90	M	75	40
Warren	−5	90	M	75	40
Williamsport	−5	90	M	74	40
York	5	90	M	75	40
RHODE ISLAND					
Block Island	5	85	L	75	40
Bristol	0	90	L	75	40
Kingston	0	85	L	75	40
Pawtucket	0	90	M	75	40
Providence	0	90	M	75	40
SOUTH CAROLINA					
Charleston	20	90	L	80	35
Columbia	20	95	M	78	35
Florence	20	95	M	79	35
Greenville	10	95	M	75	35
Spartanburg	10	95	M	78	35
SOUTH DAKOTA					
Aberdeen	−25	95	M	75	45
Huron	−20	100	H	75	45
Pierre	−20	95	M	73	45
Rapid City	−20	95	H	70	45
Sioux Falls	−20	95	H	75	45
Watertown	−25	95	M	73	45
TENNESSEE					
Chattanooga	10	95	M	76	35
Jackson	5	95	M	78	35
Johnson City	0	95	M	78	35
Knoxville	5	95	M	75	35
UTAH					
Logan	−10	95	H	65	40
Milford	−5	95	H	66	40
Ogden	−5	90	H	65	40
Salt Lake City	0	95	H	65	40
VERMONT					
Bennington	−10	90	M	73	45
Burlington	−15	90	M	73	45
Montpelier	−20	90	M	73	45
Newport	−20	85	M	73	45
Northfield	−20	90	M	73	45
Rutland	−15	90	M	73	45
VIRGINIA					
Cape Henry	15	90	L	78	35
Charlottesville	10	90	M	78	40
Danville	10	90	M	78	35
Lynchburg	10	90	M	76	35
Norfolk	15	90	L	78	35
Petersburg	10	90	M	78	35
Richmond	10	90	M	78	40
Roanoke	5	90	M	76	35
Wytheville	5	90	M	76	35
WASHINGTON					
Aberdeen	20	85	L	64	45
Bellingham	10	80	L	65	50
Everett	15	80	L	65	50
North Head	20	80	L	65	50
Olympia	15	80	M	64	45
Seattle	15	80	M	65	50
Spokane	−15	80	H	65	50
Tacoma	15	80	M	64	45
Tatoosh Island	20	80	L	65	50
Walla Walla	−10	90	H	65	45
Wenatchee	−10	90	M	65	50
Yakima	−5	90	H	67	45
WEST VIRGINIA					
Bluefield	0	95	M	75	35
Charleston	0	90	M	75	40
Elkins	−5	90	M	73	40
Fairmont	0	90	M	75	40
Huntington	0	90	M	76	40
Martinsburg	0	90	M	75	40
Parkersburg	0	90	M	75	40
Wheeling	−5	90	M	75	40

City	Winter DB	Summer DB	Daily Range	Summer WB	Latitude Deg.
Akron	–5	90	M	75	40
Cincinnati	–5	95	M	78	40
Cleveland	–5	90	M	75	40
Columbus	–5	90	M	76	40
Dayton	–5	90	M	76	40
Lima	–5	90	M	75	40
Marion	–5	90	M	75	40
Sandusky	–5	90	M	75	40
Toledo	–5	90	M	75	40
Warren	–5	90	M	75	40
Youngstown	–5	90	M	75	40
OKLAHOMA					
Ardmore	5	100	M	78	35
Bartlesville	–5	100	M	77	35
Guthrie	0	100	M	77	35
Muskogee	0	95	M	79	35
Oklahoma City	0	100	M	77	35
Tulsa	0	100	M	77	35
Waynoka	–5	105	M	75	35
TEXAS					
Abilene	5	95	M	74	30
Amarillo	0	95	H	72	35
Austin	15	100	M	78	30
Brownsville	30	95	M	80	25
Corpus Christi	25	95	M	80	30
Dallas	10	100	M	78	35
Del Rio	20	100	H	78	30
El Paso	20	100	M	69	30
Fort Worth	10	100	M	78	35
Galveston	25	95	L	80	30
Houston	20	95	M	80	30
Palestine	10	100	M	78	30
Port Arthur	20	95	M	80	30
San Antonio	20	100	M	78	30
Waco	10	100	M	78	30
WISCONSIN					
Ashland	–25	80	M	71	45
Beloit	–15	95	M	78	45
Eau Claire	–20	90	M	75	45
Green Bay	–20	90	M	73	45
La Crosse	–20	95	M	75	45
Madison	–20	90	M	75	45
Milwaukee	–15	90	M	75	45
Oshkosh	–20	90	M	75	45
Sheboygan	–20	90	M	75	45
WYOMING					
Casper	–25	90	H	62	45
Cheyenne	–20	90	H	62	40
Lander	–30	90	H	65	45
Sheridan	–30	90	H	65	45
Yellowstone Park	–35	85	H	62	45

Province & City	Winter DB	Summer DB	Daily Range	Summer WB	Latitude Deg.
ALBERTA					
Banff	–30	—	H	—	50
Camrose	–35	—	H	—	55
Calgary	–30	90	H	66	50
Cardston	–30	—	H	—	50
Edmonton	–35	90	H	68	55
Grande Prairie	–40	—	H	—	55
Hanna	–35	—	H	—	50
Jasper	–30	—	H	—	55
Lethbridge	–30	—	H	—	50
Lloydminster	–40	—	H	—	55
McMurray	–40	—	H	—	55
Medicine Hat	–35	90	H	65	50
Red Deer	–35	—	H	—	50
Taber	–35	—	H	—	50
Wetaskiwin	–35	—	H	—	55
BRITISH COLUMBIA					
Chilliwack	5	—	M	—	50
Courtenay	10	—	M	—	50
Dawson Creek	–40	—	H	—	55
Estevan Point	15	—	M	—	55
Fort Nelson	–40	—	H	—	60
Hope	0	—	M	—	50
Kamloops	–20	—	H	—	50
Kimberly	–25	—	H	—	50
Lytton	–5	—	H	—	50
Nanaimo	10	—	M	—	50
Nelson	–10	—	H	—	50
Penticton	–5	—	H	—	50
Port Alberni	10	—	M	—	50
Prince George	–30	—	H	—	55
Prince Rupert	10	—	L	—	55
Princeton	–15	—	H	—	50
Revelstoke	–25	—	H	—	50
Trail	–10	—	H	—	50
Vancouver	10	80	L	67	50
Vernon	–15	—	H	—	50
Victoria	15	—	L	—	50
Westview	10	—	M	—	50
LABRADOR					
Goose Bay	–25	—	L	—	55
MANITOBA					
Boissevain	–35	—	H	—	50
Brandon	–30	—	H	—	50
Churchill	–40	—	L	—	60
Dauphin	–35	—	M	—	50
Flin Flon	–40	—	M	—	55
Minnedosa	–35	—	H	—	50
Neepawa	–35	—	H	—	50
La Prairie	–30	—	M	—	50
Swan River	–35	—	M	—	55
The Pas	–40	—	M	—	55
Winnipeg	–30	90	M	71	50
NEW BRUNSWICK					
Bathurst	–10	—	L	—	45
Campbellton	–10	—	L	—	45
Chatham	–10	—	M	—	45
Edmunston	–15	—	M	—	45
Fredericton	–5	90	L	75	45
Moncton	–10	—	M	—	45
Saint John	–5	80	L	67	45
Woodstock	–15	—	M	—	45
NEWFOUNDLAND					
Corner Brook	0	—	L	—	50
Gander	–5	—	L	—	50
Grand Falls	–5	—	M	—	50
St. John's	0	—	L	—	50
NORTHWEST TERRITORIES					
Aklavik	–45	—	L	—	70
Fort Norman	–40	—	M	—	65
Frobisher	–50	—	L	—	—
Resolute	–40	—	L	—	—
Yellowknife	–50	—	L	—	60
NOVA SCOTIA					
Bridgewater	0	—	L	—	45
Dartmouth	0	—	L	—	45
*Halifax C	5	80	L	75	45
*Halifax A	0	80	L	75	45
Kentville	0	—	L	—	45
New Glasgow	0	—	L	—	45
Spring Hill	–5	—	M	—	45
Sydney	0	85	L	67	45
Truro	0	—	L	—	45
Yarmouth	5	—	L	—	45
ONTARIO					
Bancroft	–20	—	M	—	45
Barrie	–5	—	M	—	45
Belleville	–10	—	M	—	45
Brampton	–5	—	M	—	45
Brantford	–5	—	M	—	45
Brockville	–15	—	M	—	45
Chatham	0	—	M	—	45
Cobourg	–10	—	M	—	45
Collingwood	0	—	M	—	45
Cornwall	–15	—	M	—	45
Ear Falls	–35	—	M	—	50
Fort Frances	–30	—	M	—	50
Fort William	–25	85	M	70	50
Galt	–5	—	M	—	45
Geraldton	–35	—	M	—	50
Goderich	0	—	M	—	45
Guelph	–5	—	M	—	45
Hamilton	0	—	M	—	45
Haileybury	–25	—	M	—	50
Hanover	–5	—	M	—	45
Huntsville	–15	—	M	—	45
Kapuskasing	–30	—	M	—	50
Kenora	–35	—	M	—	50
Kingston	–10	—	M	—	45
Kirkland Lake	–25	—	M	—	50
Kitchener	–5	—	M	—	45
Lindsay	–15	—	M	—	45
London	0	—	M	—	45
Moonsonee	–35	—	M	—	50

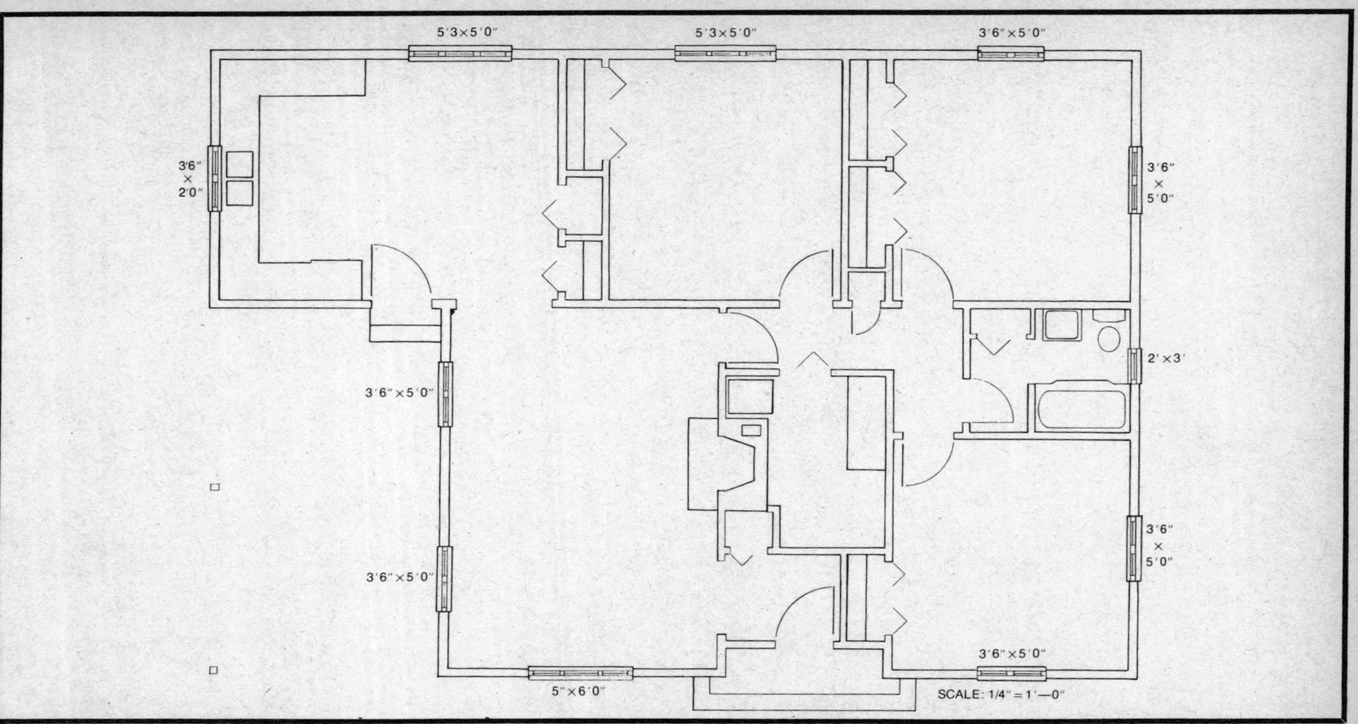

Fig. 5-2. Floor plan of a typical residence.

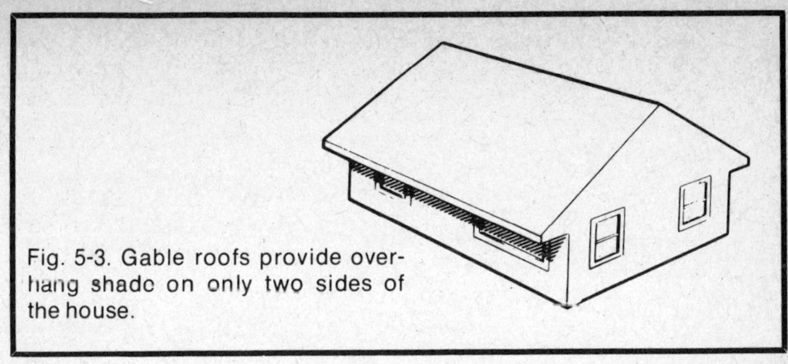
Fig. 5-3. Gable roofs provide overhang shade on only two sides of the house.

determining the number of square feet of window area on each side of the home; that is, North, South, East, and West. For example, the home in Fig. 5-2 has the following glass areas:

North Exposure:	70.5 square feet
South Exposure:	88.0 square feet
East Exposure:	41.0 square feet
West Exposure:	42.0 square feet

When measuring the area of windows for your house, the window area is the area of the wall opening in which the window is installed—not just the glass itself. Once the glass area of all exposures is found, separate the total number of square feet of glass for each side of your home into the total number of square feet of *shaded* glass and the total number of square feet of *unshaded* glass.

There are many ways a window may be shaded: awnings, tinted glass, insulating drapes; but right now we will be concerned only with shading provided by overhanging roofs. The larger the overhang, the more the walls and windows will be shaded. Also, gable roofs provide overhang shade on two sides of the house, as shown in Fig. 5-3; hip roofs provide shade on all four sides (as illustrated in Fig. 5-4). Therefore, for a

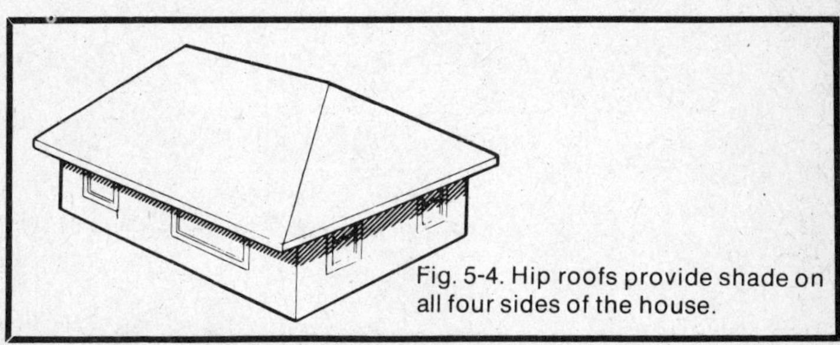
Fig. 5-4. Hip roofs provide shade on all four sides of the house.

1.	Direction which wall faces	N	S	E	W
2.	Width of Overhang, ft.	2 1	2 1	0	0
3.	Shade-Line Factor	0	2.6	.81	.81
4.	Distance Shade Line Falls below Edge of Overhang, (Line 2) x (Line 3)	0	5.2	0	0
5.	Vertical Distance from Top of Windows to Edge of Overhang, ft.		1.5	1.5	1.5
6.	Distance Shade Line Falls below Top of Windows, (Line 4) - (Line 5), ft.		3.7	0	0
7.	Total Area of Shaded Glass, sq. ft.	70.5	13.5	0	0
8.	Total Area Unshaded Glass, sq. ft. (Total Glass Area in Wall) - (Line 7)	0	74.5	41	42

Fig. 5-5. Chart for determining heat gain through shaded and unshaded glass areas.

window under an overhanging roof, figure the glass area as shaded. When no overhang exists, unshaded.

The chart in Fig. 5-5 is provided to enable the calculation of heat gain through shaded and unshaded glass areas. All dimensions entered in this chart should be in feet. If all windows in any one wall are not the same distance below the edge of the overhang or if overhang width varies along a wall, use two or more columns as necessary for that particular wall. Locate the shade line only for those windows in any wall which are shaded by an overhang.

Using the house in Fig. 5-2 as an example, we first enter the direction each wall faces and then measure the width of the roof overhang (see "A" of Fig. 5-6) for each wall. This data is entered on line 2 of the chart (Fig. 5-5), under the appropriate direction shown in line 1.

At this point we must determine the *shade line factor*—line 3 of the chart. To do this refer to the Table 5-2 and note the latitude of your location; our example house is located near Washington, D.C., so its latitude is 40°. Refer to the table below and circle the latitude that applies to your location.

Table 5-2. Shade Line Factors.

DIRECTION WINDOW FACES	LATITUDE						
	25°	30°	35°	40°	45°	50°	55°
EAST	0.83	0.83	0.82	0.81	0.80	0.79	0.78
SOUTH	10.10	5.40	3.55	2.60	2.05	1.70	1.32
WEST	0.83	0.83	0.82	0.81	0.80	0.79	0.78

Figures listed under the latitude of 40° (for our example) are the shade-line factors applying to each side of the house. These are entered on line 3 of the chart and, for our example, are 0, 2.6, 0.81, and 0.81. Note that the north side of the house has a shade-line factor of zero; the north side is considered shaded at all times and does not require a factor.

Now multiply line 2 of the chart in Fig. 5-5 by line 3 and enter your answer on line 4 as shown in our example.

Measure the vertical distance from the top of all window openings to the bottom edge of the roof overhang (shown as "B" in Fig. 5-6). These measurements are then entered on line 5 of the chart in Fig. 5-5. Subtract the figures on line 5 from line 4 and enter the answers on line 6.

Continue by multiplying the figures on line 6 by the width of the windows on each side of the house. The answer will be the total area of shaded glass and should be entered on line 7 of the chart. Our example involves only the south side of the house since the remaining sides have a zero answer. Therefore, by multiplying 1.5 (the figure on line 5) by 9 feet (total width of windows on the south side of the house) we have a total shaded area of 13.5 square feet; this figure is entered in line 7 of the chart. Since the north side of the house is shaded, we will enter thc figures (found earlier) giving the total square feet of window area on this side of the house on the chart also.

Finally, the chart is completed by subtracting line 7 from the corresponding totals of the window areas found earlier.

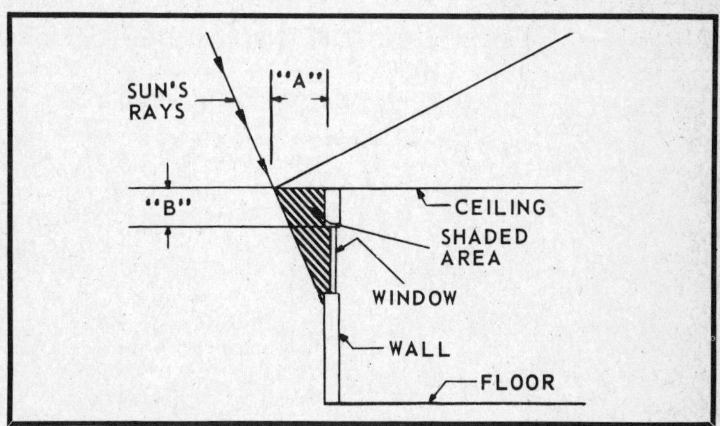

Fig. 5-6. Drawing illustrating methods of measuring window opening distances.

North exposure: 70.5 square feet – 70.5 square feet = 0 square feet

South exposure: 88.0 square feet – 13.5 square feet = 74.5 square feet

East exposure: 41 square feet – 0 = 41 square feet

West exposure: 42 square feet – 0 = 42 square feet

This gives the total area of unshaded glass and is entered on line 8 of the chart in Fig. 5-5. With this chart completed, refer again to the estimate form in Fig. 5-1 under item 1a. Here you will note three classifications of glass: regular single glass, regular double glass, and heat-absorbing double glass (tinted glass). Your windows should fall into one of these classifications. We will assume that our example has regular double glass (single glazing plus storm windows). We will then circle (2) under 1a and cross off the other two on the estimate form.

Adding up the total areas of shaded and unshaded glass from lines 7 and 8 of the chart in Fig. 5-5, we enter the total area of shaded glass on line (a); the total area of unshaded glass (east and west) on line (c); and the total area of unshaded glass (south) on line (e).

Other shade options are represented by the column headings of item 1a of the estimate form and must be considered for a correct estimate of the heat gain. Therefore, we choose the treatment applicable to our sample home, circling the factors corresponding with lines (a), (c) and (e). We will assume that our example house has drapes throughout the house, so under *Drapes or Venetian* well will circle the numbers 10, 40, and 18, respectively. Then by multiplying the window area figures entered on lines (a), (c), and (e) by the corresponding circled factor, we have:

$$54 \times 10 = 540.0$$

$$83 \times 40 = 3320.0$$

$$74.5 \times 18 = 1341.0$$

These answers are then entered in the right-hand column at the end of each line before going on to item 1b: *Air-To-Air Heat Gain.*

Begin this item by adding up the total area of all windows in the home (both shaded and unshaded) and enter the answer in the *Area Sq. Ft.* column of the Estimate Form, Item 1b, line (1) or (2) (whichever applies). Since our example has a total of 241.5 square feet of window area, this figure will be entered in the form under (2)—all double glass. The area in which our house is located has a summer outside design temperature of 90°F(DB), so this figure is circled in Item 1b under *Outdoor Design Temperature*, and the factor 4 beneath it. This factor is then multiplied by the total square feet of window area and the answer entered at the end of thc line. Our answer was 966.0.

WALLS AND DOORS

The next item (No. 2) on the Estimate Form is *Walls and Doors*, which calculates the amount of heat that enters the home through the outside walls and doors.

First, determine the total number of square feet of all outside walls and then subtract the total square feet of window area from this (since the heat gain through the windows has already been calculated). Using the residence in Fig. 5-1 as a reference, follow these steps:

1. Scale the drawings or take actual measurements of the length of each outside wall and total. In our sample home, this length totals 163.5 feet.
2. Scale or measure the floor-to-ceiling height, which is 8 feet in our example.
3. Multiply the total length (163.5 feet) by the floor-to-ceiling height (8 feet) to obtain a total of 1,308 square feet for the outside walls of the residence.
4. Subtract the total window area (the figure determined earlier and entered in item 1b, *Area Sq. ft.* column, of the Estimate Form) from the total square feet of the outside walls. Since the total window area in our residence is 241.5 square feet, subtracting this from the total wall area of 1,308 square feet leaves a net wall area of 1,066.5 square feet. This is the figure that will be used to complete Item 2 of the Estimate Form.

Below Item 2 of the Estimate Form are several classifications of wall construction: you must determine which one applies to your walls. To help you, the illustrations in Figs. 5-7 and 5-8 show various types of wall construction and insulation. Our residence has brick veneer on wood framing, so it will fall under classification (a); it also has 3 1/2 inches of

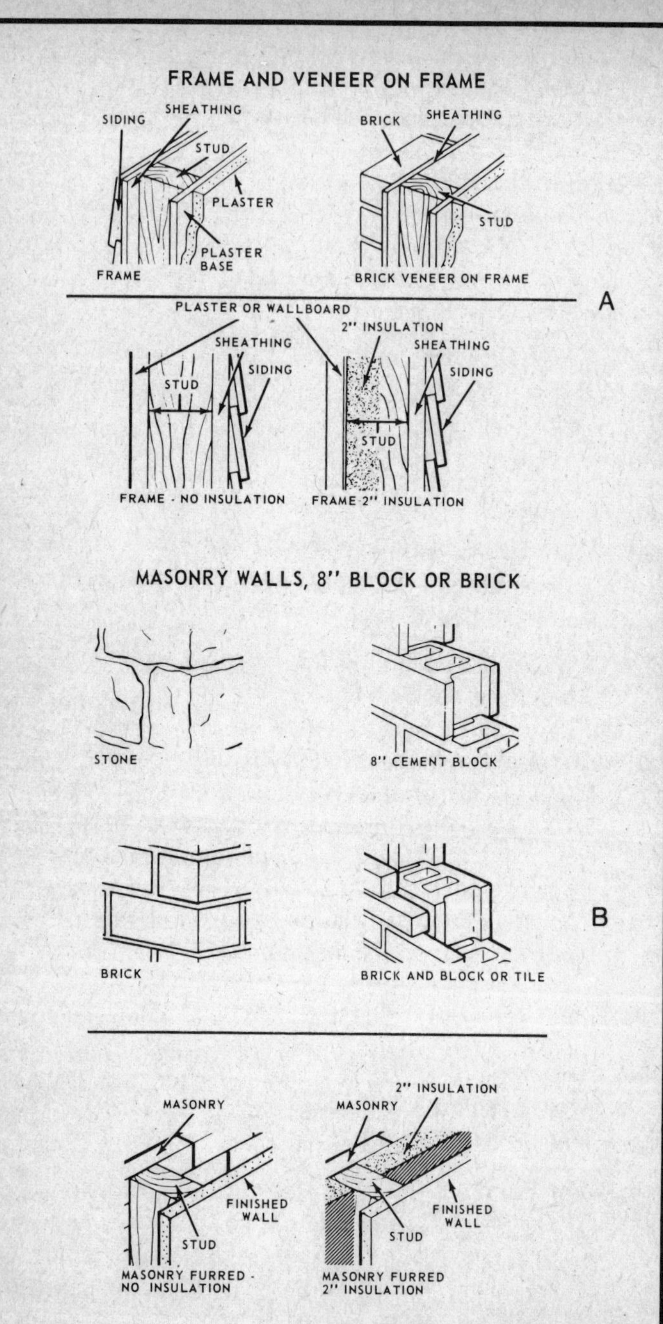

Fig. 5-7. Various types of wall construction and insulation.

insulation in the walls, so it will fall under catagory (4), *more than 2 in insulation.*

Next refer to the *Outside Temperature* and *Daily Range* entered at the top of the Estimate Form and determine the closest corresponding catagories shown in Item 2. Circle the factor beneath it on the line corresponding with the *Area Sq. ft.* figure. Since our residence is located in an area with a *medium* daily range and an outside design temperature of 90°F, we will circle the "1.5" factor on the Estimate Form. Then the net square feet of the outside walls (1,066.5) is multiplied by this factor (1.5) and the answer (1,599.75) is entered at the end of the line in the extreme right-hand column.

Below Item 2 of the Estimate Form you will note a classification (c) *Partitions* . This is for use with any partitions in your home that are not outside partitions, but are located adjacent to an area—like a garage—that is separated from the living area and is not cooled. If it does apply to your home, calculate this wall following the same procedure as for outside walls, and enter your figures on the Estimate Form. This condition does not exist in our example, so this portion of the Estimate Form will be omitted.

The last classification below Item 2 is (d) *Wood doors*. This pertains to outside doors only. In our example, we have two outside doors which are both 3 feet wide by 6 1/2 feet high; each, therefore, has an area of 19.5 square feet. The total area of these doors (39 square feet) is entered on the Estimate Form, multiplied by the appropriate factor—in the same column as you used for outside walls—and the answer is then entered at the end of the line (2d) of the Estimate Form. The answer for our example is 39 × 9.3 = 362.7.

CEILINGS

Going on to Item 3 of the Estimate Form, we can see that 3 main categories are listed: *Ceiling under naturally vented attic or vented flat roof, built-up roof no ceiling*, and *Ceilings under unconditioned rooms.* Measure your ceiling areas that apply to one (or more) of these classifications, determine the total square feet of each, and enter your figures on the appropriate line(s) of the Estimate Form. Then multiply the area by the applicable factor and enter your answer at the end of the line or lines.

The residence used as an example in this chapter has six inches of insulation in the ceiling and a light gray roof.

Customer .... Date .... Address ....

Design Conditions .... Outside temperature .... Daily Range .... Latitude ....

1a Windows—Solar Heat (See back side for example)	Area Sq. ft.	No Shades	Roller Shades	Drapes or Venetian	Ventilated Awning	Outside Shade Screens	
(1) Regular Single glass							
(a) North (or Shaded)	......	19	14	11	16	4	......
(b) Northeast and Northwest	......	51	37	28	17	12	......
(c) East and West	......	77	57	43	18	18	......
(d) Southeast and Southwest	......	66	48	36	17	9	......
(e) South	......	36	26	20	16	5	......
(2) Regular Dougle glass							
(a) North (or shaded)	54	17	13	10	11	3	540
(b) Northeast and Northwest	......	45	35	26	12	8	......
(c) East and West	83	67	53	40	12	12	3320
(d) Southeast and Southwest	......	57	45	34	12	6	......
(e) South	74.5	31	25	18	11	3	1341
(3) Heat Absorbing Double glass							
(a) North (or shaded)	......	10	8	7	8	2	......
(b) Northeast and Northwest	......	26	22	18	9	7	......
(c) East and West	......	40	33	28	10	10	......
(d) Southeast and Southwest	......	34	28	23	9	5	......
(e) South	......	20	16	13	9	3	......

1b Windows—air to air heat gain	Area Sq. ft	Outdoor Design Temperature					
		90	95	100	105	110	
(1) all single glass		8	12	16	20	24	
(2) all double glass	241.5	4	7	9	11	13	966

2 Walls and Doors	Area Sq. ft	Daily Temperature Range										
		Low		Medium				High				
		Outdoor Design Temperature										
		90	95	90	95	100	105	95	100	105	110	
a Frame and Veneer on frame												
(1) No insulation	......	5.9	7.2	4.8	6.1	7.4	8.7	4.8	6.1	7.4	8.7	......
(2) less than 1 in. insulation	......	4.3	5.2	3.5	4.5	5.4	6.4	3.5	4.5	5.4	6.4	......
(3) 1 to 2 in. insulation	......	2.9	3.6	2.4	3.1	3.7	4.4	2.4	3.1	3.7	4.4	......
(4) more than 2 in. insulation	1,066.5	1.8	2.2	1.5	1.9	2.3	2.7	1.5	1.9	2.3	2.7	1,599.75
b Masonry walls 8 in. block or brick												
(1) plastered or plain	......	7.3	9.7	5.4	7.8	10.2	12.6	5.4	7.8	10.2	12.6	......
(2) Furred, no insulation	......	4.6	6.1	3.4	4.9	6.4	7.9	3.4	4.9	6.4	7.9	......
(3) Furred, less than 1 in. insulation	......	3.1	4.1	2.3	3.3	4.3	5.3	2.3	3.3	4.3	5.3	......
(4) Furred, 1 to 2 in. insulation	......	2.1	2.8	1.6	2.3	3.0	3.7	1.6	2.3	3.0	3.7	......
(5) Furred, more than 2 in. insulation	......	1.4	1.8	1.0	1.5	1.9	2.4	1.0	1.5	1.9	2.4	......
c Partitions												
(1) Frame, finished one side only, no insulation	......	8.4	11.4	6.0	9.0	12.0	15.0	6.0	9.0	12.0	15.0	......
(2) Frame, finished both sides, no insulation	......	4.8	6.5	3.4	5.1	6.8	8.5	3.4	5.1	6.8	8.5	......
(3) Frame, finished both sides, more than 1 in insulation	......	2.0	2.7	1.4	2.1	2.8	3.5	1.4	2.1	2.8	3.5	......
(4) Masonry, finished one side, no insulation	......	2.6	4.4	1.2	3.0	4.7	6.5	1.2	3.0	4.7	6.5	......
d Wood doors*	......	11.3	13.8	9.3	11.8	14.3	16.8	9.3	11.8	14.3	16.8	......

Item		Quantity											Btuh
3 Ceilings and Roofs													
a Ceiling under naturally vented attic or vented flat roof													
(1) Uninsulated	dark	......	9.9	11.0	9.0	10.1	11.3	12.4	9.0	10.1	11.3	12.4	......
	light	......	8.1	9.2	7.1	8.3	9.4	10.6	7.1	8.3	9.4	10.6	......
(2) Less than 2 in. insulation	dark	......	4.3	4.8	3.9	4.4	4.9	5.4	3.9	4.4	4.9	5.4	......
	light	......	3.5	4.0	3.1	3.6	4.1	4.6	3.1	3.6	4.1	4.6	......
(3) 2 in. to 4 in. insulation	dark	......	2.6	2.9	2.3	2.6	2.9	3.2	2.3	2.6	2.9	3.2	......
	light	......	2.1	2.4	1.9	2.2	2.5	2.8	1.9	2.2	2.5	2.8	......
(4) More than 4 in. insulation	dark	......	1.7	1.9	1.6	1.8	2.0	2.2	1.6	1.8	2.0	2.2	......
	light	......	1.4	1.6	(1.2)	1.4	1.6	1.8	1.2	1.4	1.6	1.8	1,628.4
b Built up roof no ceiling													
(1) Uninsulated	dark	......	17.2	19.2	15.6	17.6	19.6	21.6	15.6	17.6	19.6	21.6	......
	light	......	14.0	16.0	12.4	14.4	16.4	18.4	12.4	14.4	16.4	18.4	......
(2) 2 in. roof insulation	dark	......	8.6	9.6	7.8	8.8	9.8	10.8	7.8	8.8	9.8	10.8	......
	light	......	7.0	8.0	6.2	7.2	8.2	9.2	6.2	7.2	8.2	9.2	......
(3) 3 in. roof insulation	dark	......	6.0	6.7	5.5	6.2	6.9	7.6	5.5	6.2	6.9	7.6	......
	light	......	4.9	5.6	4.3	5.0	5.7	6.4	4.3	5.0	5.7	6.4	......
c Ceilings under unconditioned rooms		......	2.7	3.6	1.9	2.9	3.8	4.8	1.9	2.9	3.8	4.8	......
4 Floors													
a Over unconditioned rooms		......	3.4	4.6	2.4	3.6	4.8	6.0	2.4	3.6	4.8	6.0	......
(b) Over basements, enclosed crawl space, or slab on ground		......	0	0	(0)	0	0	0	0	0	0	0	0
c Over open crawl space		......	4.8	6.5	3.4	5.1	6.8	8.5	3.4	5.1	6.8	8.5	......
5 Outside Air													
a Infiltration, Btuh per sq ft of gross exposed wall area		1308	1.1	1.5	(1.1)	1.5	1.9	2.2	1.6	1.9	2.2	2.6	1,438.8
b Mech. ventilation Btuh per cfm		370	13.2	21.6	(16.2)	21.6	27.0	32.4	21.6	27.0	32.4	37.8	5994
6 People (never less than 3)			4 Number of occupants x 300 Btuh										1200
7 Kitchen appliance allowances			1200 Btuh										1200
SUB TOTAL			Items 1 thru 7										19,227.4
8 Heat gain to ducts			Subtotal Item 1 thru 7 x factor**										0
9 Total Estimated Sensible Heat Gain			Total (Items 1 thru 8)										19,227.4
10 Latent heat gain allowance			Item 9 x 0.3										...576.83
11 Total Estimated Design Heat Gain			Item 9 + Item 10 =										19,804.23

*Consider glass area of doors as windows. **Factor for ducts in attic with 2 in. insulation .10 Factor for ducts in attic with 1 in. insulation .15; For ducts in slab floors, insulated ducts in enclosed crawl space, unconditioned basement or furred spaces use factor of .05.

Fig. 5-8. Completed estimate form for the sample residence in this chapter.

Therefore, we will circle the appropriate factor under (a), classification (4). Referring to the proper column for our outdoor design conditions, the factor for our example is 1.2. Multiplying this by the total ceiling area (1,357), we obtain an answer of 1,628.4 to be entered at the end of line 4 (light) under classification (3a) on the Estimate Form.

FLOORS

Item 4, *floors*, is very similar to the calculations for the ceilings and roofs in Item 3. Again there are 3 classifications: (a) over unconditioned rooms; (b) over basements, enclosed crawl space, or slab on ground; and (c) over enclosed crawl space. Determine the classification of your floors from the lines, calculate the area in square feet and enter your answer on the appropriate line(s) of the Estimate Form. Multiply this area by the applicable factor in the column under your outside design conditions and enter your answer at the end of the line or lines.

Our residence has an enclosed crawl space ("B" under 4), so the factor will be zero—no heat gain through the floor. Therefore, zero will be entered at the end of the line.

Refer to Item 2 of the Estimate Form—*walls and doors*—to obtain the total area of the walls and doors; then add the total area of the windows to this figure in order to obtain the gross area of all outside walls. Enter this figure on line (a) under Item 5...1,308 square feet for our sample residence. Multiply this figure by the applicable factor (1.1 in our case) and enter the answer 1,438.8 at the end of line (a).

Mechanical ventilation or exhaust fans are used during cooking or bathing and therefore must be considered in calculating your air-conditioning system. Check the nameplates of all fans in your home to obtain the CFM rating of each. As a rule of thumb, most kitchen/range exhaust fans will average about 300CFM, while bath fans will be around 70 CFM. The residence used as an example has one kitchen fan at 300 CFM and a bath fan at 70 CFM for a total of 370 CFM. This figure is entered in the *Area Sq. ft.* column and then multiplied by the applicable factor (16.2) for a total of 5,994, which is entered at the end of line (b) under Item 5.

Continuing on to Item 6—*People*—enter the number of people normally occupying the home, but never less than 3. We will assume 4 for our residence; this number is multiplied by

300 (Btuh) for an answer of 1200 to be entered at the end of the line.

Unless you have an extremely large kitchen in your home where lots of heat-producing appliances will be operating at the same time, use the figure 1200 at the end of line 7. Consult your local utility company if you do have many appliances operating at once.

Now total Item 1 through 7 to obtain a subtotal of heat gain in the home; your answer will be in Btuh. Our example comes out to 19,227.4 Btuh.

If you have concealed heating ductwork (such as is found when ductwork is installed beneath an existing ceiling and a new lowered ceiling is built in to conceal it), multiply the *subtotal* figure by 0.05 and enter your answer at the end of the line. Since this condition does not exist in our residence, this step has been omitted.

To complete the Estimate Form, we total lines 1 through 8 (19,227.4 in our case) and enter this answer on line 9. Then multiply this figure by 0.3 to obtain an allowance for latent or hidden heat. The calculations will be complete when Item 9 and Item 10 are added and the answer entered on line 11.

The completed Estimate Form used to calculate the heat gain for the sample residence is shown in Fig. 5-8. We now have accurately determined the amount of heat that enters the home on a summer day when the outside temperature reaches the outdoor design temperature. Now we are ready to select cooling equipment for the home.

If individual room or window air-conditioning units are to be used, we will have to break the total load down into areas. This is done simply by determining the number of Btuh per square foot and then installing cooling units equal to the total square feet in the area to be cooled multiplied by the Btuh-per-square-foot factor. For example, our residence has a heat gain of 19,804.23 Btuh for the entire house. This means that 14.59 Btuh of cooling will be required for each square foot of living space. Therefore, if we wanted to cool, say, just the master bedroom (156.25 square feet), we multiply 156.25 by 14.59 for a total of 2,279 Btuh of cooling required in this area. The remaining areas may be calculated in the same way.

Of course, if a central unit is desired, you will use one capable of offsetting the heat gain determined on the Estimate Form. If you are planning to use an add-on air-conditioning system like the one described in Chapter 7, there are a few other items you should look into before deciding upon the equipment to be used in your particular system. You must first determine what type of system you presently have and how well it will accommodate the addition of central air conditioning.

Investigate your present heating system and its related ductwork to determine the size of the furnace blower, the number and sizes of the branch ducts, and the type of grilles and register. Also check your electrical system to make certain that you have enough spare capacity to handle the additional load of an air-conditioning system—you may have to have some additional wiring done before you install the system.

McGraw-Edison dealers have a printed form for analyzing your home's existing heating system that gives step-by-step procedures; all you'll need to complete the survey is a pencil, a flashlight, a measuring tape, and some scratch paper for figuring and making notes. If you have trouble determining your heat gain, the dealer will also be glad to give you assistance with this.

Chapter 6

Installing Combination Heating & Cooling Units

Everyone's goal, as far as heating and cooling the home are concerned, is comfort at low cost. There are many ways to achieve this goal, but one easy way is to use thru-wall heating and cooling units in various areas of the household. Such a system gives the occupants complete control of their environment with a room-by-room choice of either heating or cooling at any time of year—at any temperature they desire.

These units are easily installed in new construction or during modernization, and comfort conditioning can then be controlled floor-by-floor, wing-by-wing, or room-by-room. The initial cost of such a system usually is much less than a central air-conditioning system offering comparable performance. Add a normal life expectancy of 20 years or more and you have a heating and cooling system that is hard to beat.

Operating costs are lower than many other systems due to the high efficiency of room-by-room control. The living area, for example, can be heated at a temperature of 70°F, while the bedrooms may require only 55°F. Other areas which are used only occasionally—like a guest bedroom or workshop—may be kept at an even lower temperature until occupied. The same principle applies to cooling the areas during warm months. Unlike a central forced-air duct system, the thru-wall units never need balancing from season to season and, if a unit should fail, the defective chassis can be replaced immediately or taken to a shop and repaired.

But don't pick up your phone and order a dozen or so just yet—this type of comfort conditioning may not be the best for your needs. First, at each location, besides having the relatively large unit protruding inside your home, there's an opening, say 15 inches × 54 inches, on the outside of your home; this is the grille for air-cooling your air-conditioning compressor and for make-up air. Will these openings, in any way, damage the esthetics of your home?

Another consideration is the noise that these units make due to the compressor and blower fan. The better units are very quiet, but some on the market are so noisy that you would not want them in your home. Therefore, it seems in order for you to examine several brands closely before making a decision.

The esthetics of these thru-wall units can be improved somewhat by making certain that the opening is cut to fit the sleeve and louvered grille correctly. Also make sure that the grille fits flush with the outside finished surface of your home. The grilles and trim may then be painted a color to match the finish of your home, making them hardly noticeable. Low shrubs planted in front of the grilles will make them even less conspicuous, but you must provide enough clearance for intake and exhaust air.

If, after having read Chapter 3 of this book and considering all factors involved, you still want to install thru-wall units in your home, the following paragraphs will show you step by step how to select and install this type of system. Of course, exact dimensions and procedures will vary from manufacturer to manufacturer, so always study the installation instructions accompanying your unit.

SIZING THE UNITS

When planning thru-wall heating and cooling units for your home, you will have to make room-by-room heating and cooling calculations, following the procedures given in Chapters 2 and 5 of this book. Make these calculations very carefully as an improperly sized unit (over or under) will cost you in several ways:

1. Higher equipment costs.
2. Money wasted if unit has to be replaced with one of the correct size later on.
3. Higher operating costs.
4. Expected comfort will not be realized.

Once you have made a survey of your home and made the various heating and cooling calculations on the forms included in this book, it wouldn't be a bad idea to have an engineer at your local utility company check over your work. Most companies will be glad to do this for you at little or no charge. However, if you experience difficulty in getting your calculations checked by the utility company, the equipment manufacturers will most certainly do this for you—especially if you're buying the equipment from them.

LAYING OUT THE UNITS

Let's assume that the floor plan of the residence in Fig. 6-1 is similar to the one in which you install a thru-wall heating and cooling unit. We have already made the heating and cooling calculations and find them to be as follows:

AREA	HEAT LOSS (for heating)	HEAT GAIN (for cooling)
Living room	12,619 Btuh	7140 Btuh
Kitchen	11,222 Btuh	6733 Btuh
Utility Room	1670 Btuh	No cooling
Bath	2170 Btuh	No cooling
Bedroom No. 1	6945 Btuh	4167 Btuh
Bedroom No. 2	7384 Btuh	4430 Btuh
Bedroom No. 3	6230 Btuh	3738 Btuh

From these listings, we see that since the utility room and bath will not be cooled, we will need five thru-wall heating and cooling units—as close to the ratings given as possible—to comfort-condition the house. Some other type of heat will be supplied for the bath and utility room. Before ordering the units, however, you should obtain shop drawings of several different types of units in order to check the rough-in dimensions against your house structure. Perhaps you want to mount the units under existing windows; one brand may be too high to fit, while another brand may fit just right. Check out and plan the system carefully before ordering any equipment and especially before doing any cutting on your walls.

Once you have ascertained that the units can be made to fit all existing conditions, your next step is to check the capacity of your electric service and make sure that routing circuits to each unit is feasible without too much cutting and patching.

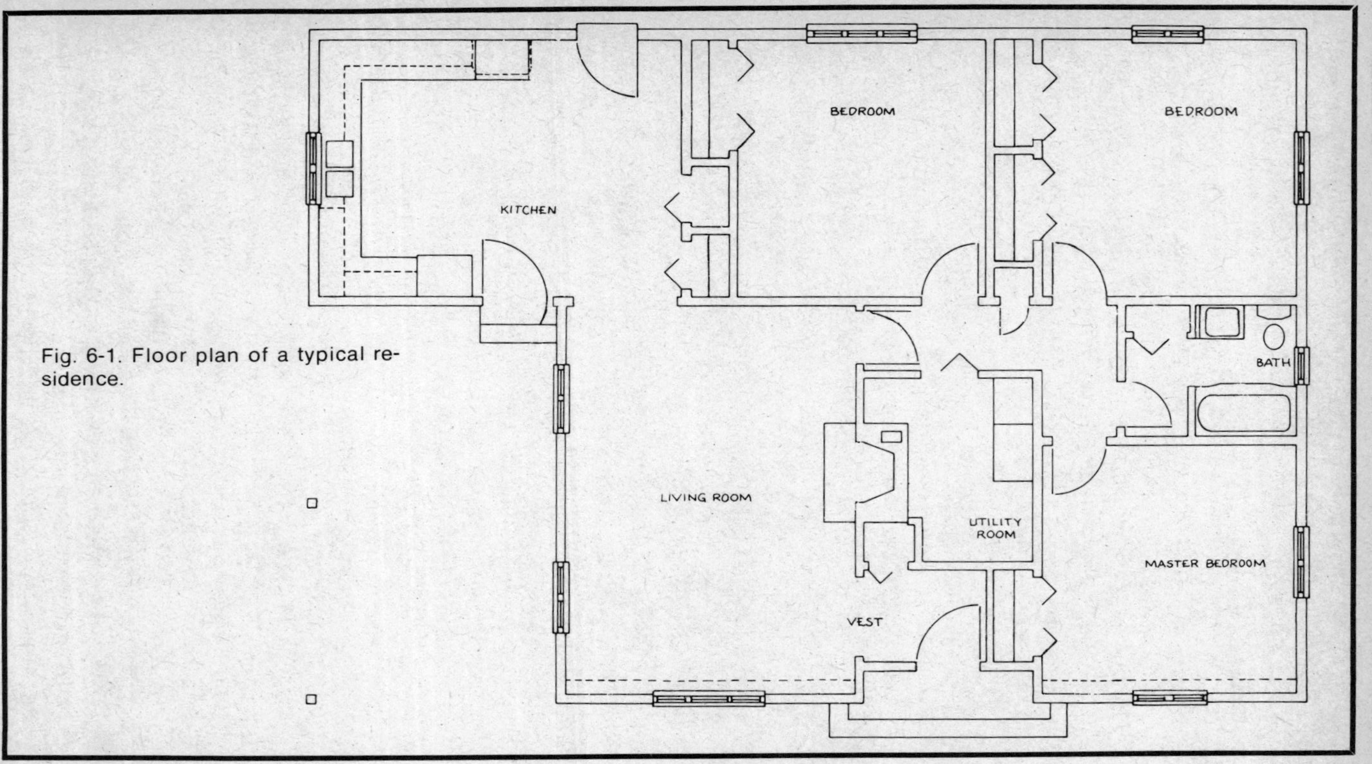

Fig. 6-1. Floor plan of a typical residence.

You may want to have an electrical contractor give you an estimate on the wiring of the units, or you can install the circuits and make the connections yourself by following the instructions given in Chapter 4.

Like other types of heating equipment, the thru-wall units should be located on an outside wall near the areas where the greatest heat loss will occur, such as under windows, etc. Most thru-wall units have built-in controls designed to sense the return air and ambient temperatures, so you won't have to bother with installing separate controls when installing these units.

SELECTING AND ORDERING THE UNITS

You should have gathered plenty of literature on the various models of thru-wall units available when you were checking their dimensions against those in your home. Using this literature, select a model that will closely match the heat loss and heat gain estimates obtained in your calculations. You will probably never hit the figures exactly, but come as close as you can. In general, if you can come within 10% of your heat-loss figures and within 15–20% of your heat-gain figures, you will be close enough for good comfort conditioning. However, never drop below these percentage margins, especially on the heating equipment.

In most cases you're going to get what you pay for, but shop around with different suppliers before ordering the equipment. Buying the units through a contractor friend of yours may also save some money.

Delivery of these units may take some time, so once you have decided on the brand, size, and supplier, order them several weeks before you're going to need them. While waiting you could go ahead and rough-in the electric wiring and change your electric service (if this is necessary to accommodate the additional load). You may even want to add a little more insulation to help keep your operating costs to a minimum, or perhaps install some new storm windows for the same reason.

INSTALLATION INSTRUCTIONS

When your units arrive, each will consist of four basic components as shown in Fig. 6-2: (1) base heater, including room-size fans, heating element, and filter; (2) outdoor

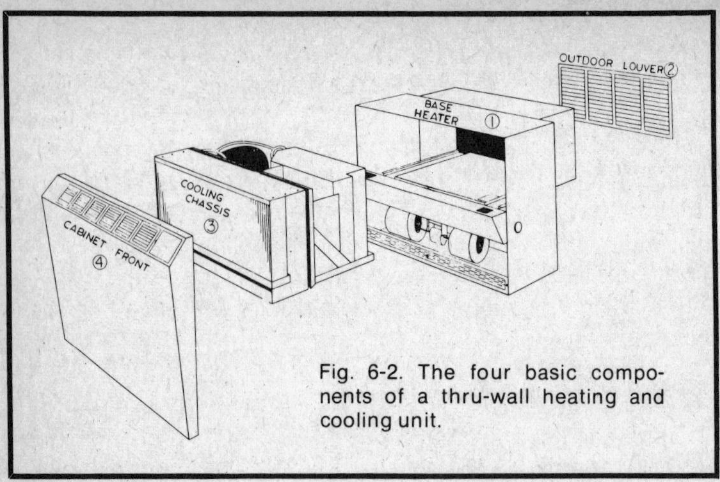

Fig. 6-2. The four basic components of a thru-wall heating and cooling unit.

louver; (3) cooling chassis; and (4) cabinet front. In most cases these components are all that is necessary for a complete installation other than some hand tools, a few miscellaneous items found at your local hardware store, and a little elbow grease.

Before starting the installation, review your wall thicknesses and make certain that your measurements have been accurate as they will dictate the method of installing the base unit. Where the walls are thicker than 9 1/4 inches, or where built-in radiator enclosures are desired or already provided, usable floor space can be saved and the appearance of the installation enhanced if the base cabinet is partially recessed in the wall. However, if your walls are less than 9 1/4 inches thick, the rear extension may project beyond the outside face of the wall as shown in Fig. 6-3A. Obviously, this is undesirable, and the unit should be made flush with the outside wall by using a filler piece as illustrated in Fig. 6-3B. If it is necessary to use this filler strip, add studs between the cabinet and the wall so the cabinet can be fastened securely.

Start your wall openings following the rough-in dimensions packed with your unit. If you are cutting through wooden studs of an existing building, frame the opening before installing the sleeve through the wall. Most units are designed to fit exactly in a given number of block or brick courses so, if you're installing in a masonry wall, all you'll probably need to do is remove the required number of blocks or bricks and insert the sleeve. Once the unit is in place the equipment itself will provide all the support necessary, so don't worry about angle-iron lintels.

Dimensions for locating the wall opening for typical thru-wall heating and cooling units are shown in Fig. 6-4. Figure 6-5 gives useful suggestions for cutting the wall opening and sealing it once the sleeve is in place. The opening should have a neat finish, be weatherproofed, and the bottom of the opening should be pitched towards the outside approximately 1/4 inch per foot for drainage.

At this time, you should also make preparations for the electrical wiring and—if you're using hot-water heat—for water pipes.

Before installing the base section and rear extension, apply non-hardening caulking compound around the inside of the wall opening as described in Fig. 6-5. Then apply caulking compound around the outside of the base rear extension, to about one inch from the rear edge of the cabinet body. Caulk the top and sides only—at this time—unless the bottom edge of the masonry is rough. If this is the case, the bottom edge may be caulked also, but be sure to leave two or three one-inch gaps (weep holes) in this caulking for moisture drainage.

Next place the base heater rear extension in the wall opening and slide it to the desired location. Shim the bottom of

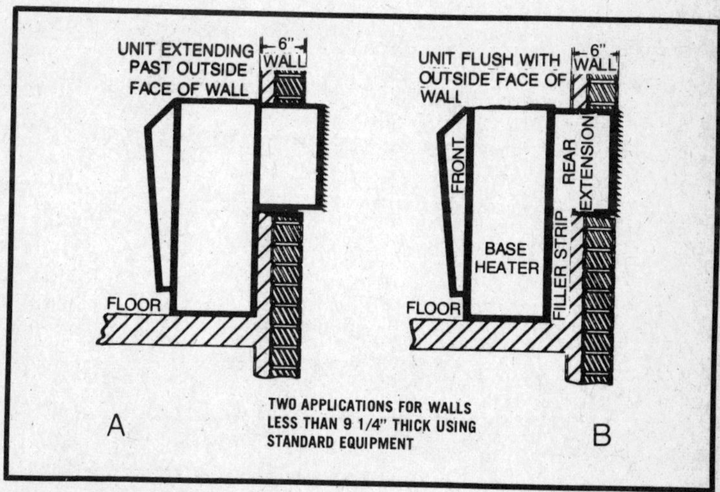

Fig. 6-3A. On thin walls, the rear extension may project beyond the outside face of the wall, giving an undesirable appearance. Fig. 6-3B. The projection is solved by using a filler piece to make the unit flush with the outside wall.

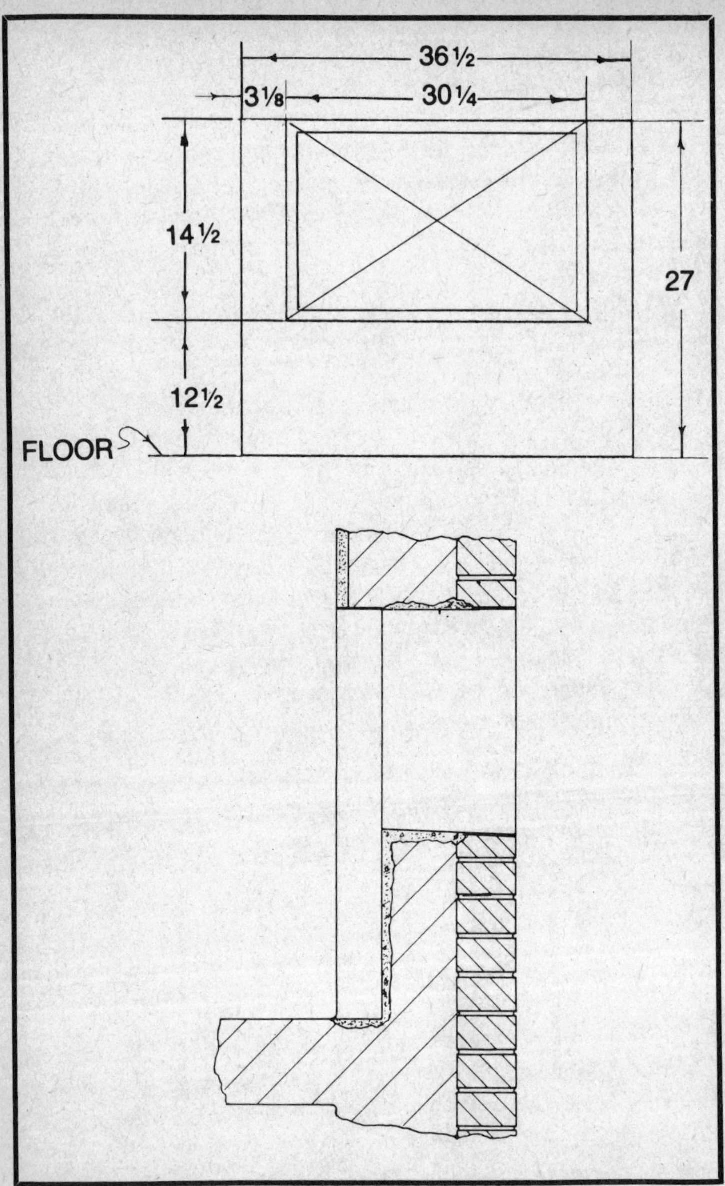

Fig. 6-4. Wall section and dimensions for installing a recessed-cabinet unit.

the cabinet, if necessary, to make the rear extension approximately 1/4 inch low (below horizontal) at the end nearest the outside face of the wall. This is important as it prevents condensed moisture from running toward the inside of your house.

Secure the base unit to the floor and wall using suitable fasteners and check the caulking around the entire rear extension at the outside face of the wall. If the bottom face of the wall has been caulked, check the weep holes to make certain they have not been clogged by the caulking.

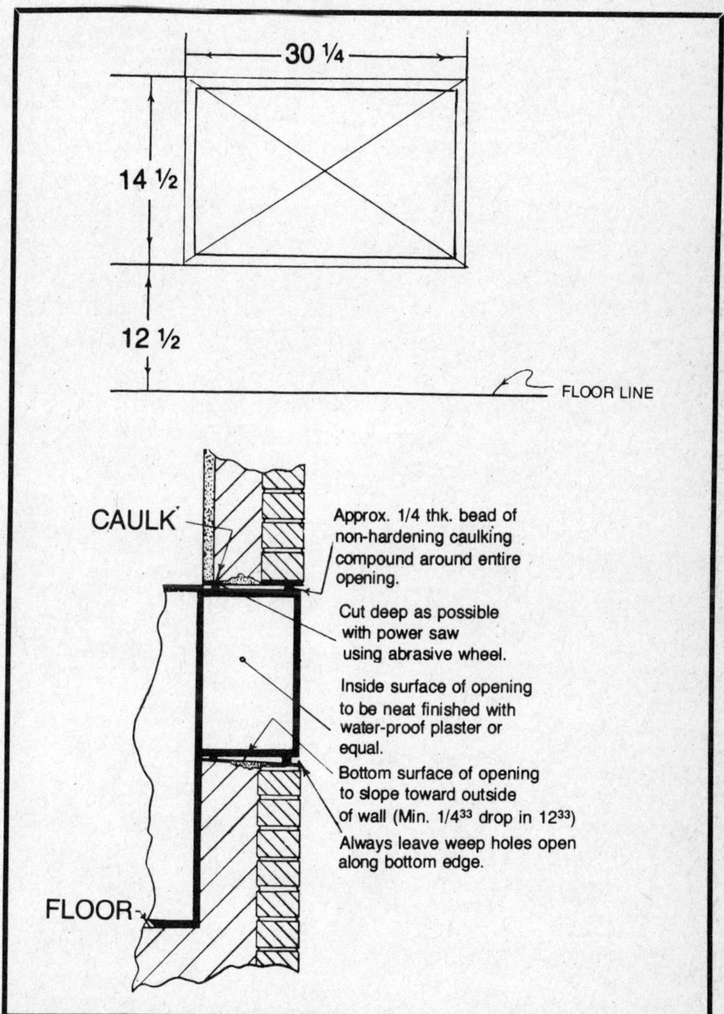

Fig. 6-5. Wall sections and dimensions for installing a typical non-recessed unit. Finishing and cutting instructions also apply to recessed units.

Now we come to a fork in the road; the one to take depends entirely upon the type of fuel you are using. You could be replacing old steam or hot-water radiators. In this case, you may want to use a thru-wall unit with a steam or hot-water heating coil; the existing pipes can be used to supply the coil. On the other hand, your old heating system may not have used pipes, in which case the simplest course is to use electric thru-wall units. Therefore, we will describe the procedures for installing both types at this point. Remember, however, that the cooling cycle of both types requires electricity.

Piping For Steam Or Hot Water

The steam or water-heating coil in most thru-wall units may be removed in order to make connections by removing two bolts at each end of the coil. However, the coil should be securely bolted in place and properly pitched before the final pipe connections are made. Either end of the coil may be used as inlet or return. The use of 1/2-inch I.D. (5/8-inch O.D.) copper tubing is recommended for connecting the heating coil to the pipe.

Once the heating coil has been pitched and bolted securely in place, the pipe connections may be completed as shown in Fig. 6-6. Note that holes are provided in both the back and bottom of the cabinet for heat piping. If you cannot arrange your existing piping to utilize these holes, others may be drilled in the cabinet to suit your particular needs.

For the type of unit shown in Fig. 6-6, you will need two 3/4-inch × 2 1/2-inch black iron nipples (item 3), one 3/4-inch I.P.S. × 5/8-inch O.D. female copper adapter, and two lengths of 5/8-inch soft copper tubing long enough to reach from the heating coil to your existing piping. A shut-off valve is recommended on the inlet pipe to the coil, mounted as shown in Fig. 6-6 (item 8). In the case of hot-water piping, be sure to include an air vent (item 26). Other items will include a 1/2-inch balancing fitting (item 9) and one 1/2-inch × 1/2-inch × 3/4-inch I.P.S. iron reducing tee. These items are not normally furnished with the units and will have to be purchased at your local hardware store or plumbing supply house.

With the piping out of the way, slide the blower assembly back into the cabinet extension (if you had to remove it) and reconnect the wires to the thermostat and junction box. However, do not install the mounting plate at this time.

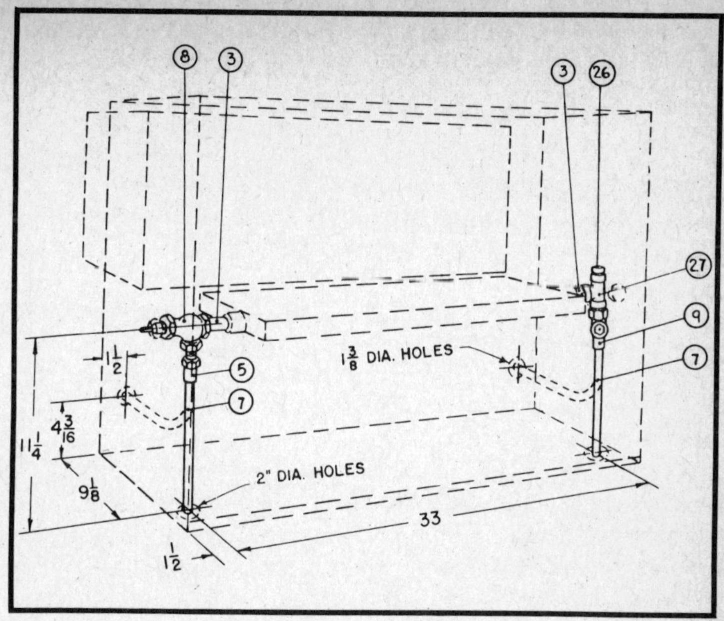

Fig. 6-6. Pipe connections for a typical thru-wall heating and cooling unit.

You're now ready to install the cooling chassis, but it's going to require some help as it weighs too much for one man to handle. When you've rounded up that neighbor (and both of you have finished your brew), uncrate the cooling chassis as close as possible to the base unit in which it is to be installed. With your hand, spin the condenser blower wheel to be sure that it has not become loose due to rough handling in shipment. If it has been loosened or the wheels rub against the blower housing, center the set screw over the flat part of the motor shaft and retighten it.

Slide the cooling chassis all the way into the base section, but be extremely careful not to lift or pull it by any of the copper tubing forming the refrigeration circuit. When properly in place, reinstall the mounting plate.

Before proceeding any further, take another break (you'll probably need it anyway after lifting the heavy cooling chassis in place) and inspect the base convector rear extension to make sure it is sealed and insulated properly. If any light shows through, or if there is any way air might leak between the base section and the wall, caulk and insulate as necessary. This is very important since gaps will allow unconditioned outside air to infiltrate. In addition, in extremely cold weather gaps could cause steam traps or hot water lines to freeze.

Electrical Wiring

A separate electrical circuit should be used for each conditioner; each should be provided with a disconnect device at the unit location. The feeder wire and over-current protection should be sized in accordance with the National Electrical Code.

Always check the nameplate ratings on the cooling chassis (and on the heating coil if you're using an electric coil). In nearly all cases, the supply voltage will be 240 volts, but the amperage rating will vary with the size of the unit. If electric heat is used, the heating element load will be larger than the cooling load; since both operations will never run simultaneously, the wire size and overcurrent protection may be sized for the larger of the two, not for their sum.

For example, if the nameplate rating on your unit is 7.5 amps at 230 volts for the cooling chassis and you are using hot water for heat, 7.5 amps is the total load. Therefore, the wire would be sized as follows:

$$7.5 \text{ (amperes)} \times 1.25 \text{ (safety factor)} = 9.37 \text{ amperes}$$

Referring to the wire table (Table 6-1), we find that No. 14 AWG is the smallest wire listed and is rated to carry 15 amperes. This is the size of wire to use. Conventional circuit breakers are not rated smaller than 15 amperes, so you will use a 15-ampere breaker.

But suppose the nameplate stated an amperage of 14.9 amperes. Again, applying the safety factor we have $14.9 \times 1.25 = 18.6$ amperes. Checking in the table, we find that we now have to go to No. 12 AWG wire and, in turn, provide an overcurrent device rated at 20 amperes.

Table 6-1.
Current-Carrying Capacity of Copper Wire.

WIRE SIZE	AMPERE RATING
14	15
12	20
10	30
8	40
6	55

If you use an all-electric thru-wall unit, a typical nameplate rating could state:

Cooling Amperes—7.5
Heating Amperes—12.5

In this case, you would size the wire and overcurrent protection for the larger of the two loads—the heating load.

Besides the overcurrent protection at your electric panel, you will need a separate disconnect device at the unit itself. This can be accomplished in several ways:

1. Have a remote switch in the power line to the unit.
2. Install a double-pole, single-throw (ON-OFF) switch of the proper rating in the cover of the junction box provided.
3. Install a 240-volt receptacle in the junction box and attach a plug to the unit's power cord.

In all cases, be absolutely certain to ground the base cabinet to guard against harmful electric shock should a fault occur in the wiring. If armored cable is used for the circuit, the connection to the chassis will also provide a ground, provided the cable is properly grounded at the breaker panel. If non-metallic cable with ground wire is used, the bare ground wire should be securely fastened under some screw in the chassis or base section. Most units will have a special grounding screw (painted green) for this purpose.

With the wiring completed, install the filter and the cabinet front. Press the control button marked OFF, and turn on the electric power supply. If using a steam or hot-water heating coil, make sure the inlet valve is turned off. Then press the button marked COOL, and turn the thermostat knob clockwise to the extreme COOLER position. The compressor will start immediately if the heating coil is cool, and in a few minutes the unit should be discharging cooled air.

The compressor may not start if the room is very cold; it may be necessary to submerge the thermostat bulb in warm water to get the compressor to start. Now turn the thermostat knob to the extreme WARMER position; the compressor should stop unless the room is very warm.

Now let's test the heating cycle. Press the button marked OFF and wait a few minutes before pressing the one marked HEAT, leaving the thermostat at the extreme WARMER position. The circulating-air fans will run at low speed—for normal heating—if the room is not too hot.

If you install an all-electric model, the heating element will also be energized and warm air will be felt in a few seconds. If the unit uses steam or hot water, you won't feel any significant change in temperature of the air unless your furnace and circulating pumps are operating. But if the blower fan operates, you know the unit itself is working properly. Press the button marked OFF and all motors should stop. You have now completed your first unit.

Continue this procedure for the remaining units until all are installed. Then you will probably want to paint the outside louvers to match the finish of your house. The inside unit can also be painted to match any room decor, including wood-grain finishes, or you might prefer to build a wooden enclosure around the unit, leaving an opening for the air discharge and return.

Chapter 7
Central Air-Conditioning Units

Central air conditioning was considered a prohibitive luxury for most homeowners only a few years ago. They had to be content with a heating system and perhaps a window air conditioner in the bedroom or family room. But central air conditioning now comes in kit form with costs cut to a bare minimum. The average handyman can make the installation in an eight-room house (that has forced-air heat with an adequate blower and ductwork) for under $1200.

The working principle is simple (see Fig. 7-1). You install a compact cooling coil in the plenum (warm-air outlet) of your furnace and a remote condensing unit outside your home, but as close as possible to the furnace; refrigerant lines connect the condensing unit to the cooling coil. The furnace blower pushes air through the coil and the existing ductwork. In winter, the furnace runs as usual with the condensing unit turned off. The cooling coil has a negligible effect on the volume of heated air coming from the furnace.

Several companies manufacture central air-conditioning systems designed especially for the do-it-yourselfer. Most of these kits require little skill and only a few hand tools. Since the cooling coil is designed to fit in your existing forced-air furnace and uses the heating ductwork for distribution, the average installation will take only a single day to complete.

The cost of the kits ranges from $700 for a two-ton unit to over $1000 for a four-ton (48,000 Btuh) unit. The length of the

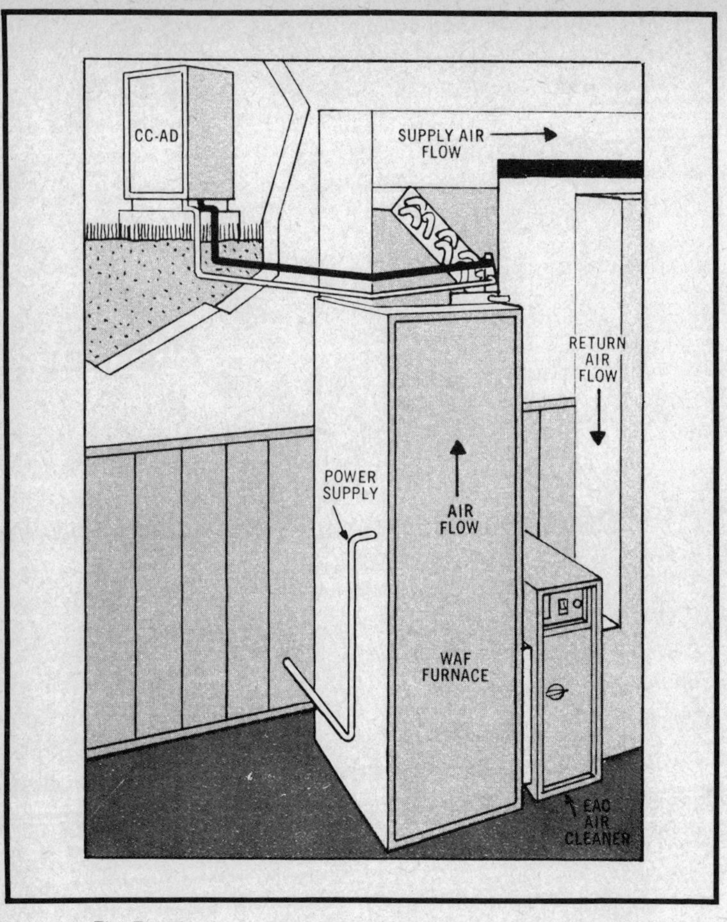

Fig. 7-1. Basic components of an add-on cooling system.

refrigerant tubing, type of controls, and, if necessary, a replacement blower for your furnace will affect your total cost. An electrician will probably charge around $100 to install the wiring required for the condensing unit. Add another $75 for a concrete pad for the condensing unit and we have a total cost of between $1000 and $1500 for the complete installation.

In general, there are two basic steps to adding central air conditioning to your existing forced-air heating system:

- An analysis of your home's cooling requirements.
- Selecting and installing the system.

A careful analysis of your home's cooling requirements is an all-important step to help you select the proper size and type of equipment needed for your particular home and forced-air system. It is important because an oversized unit can cause as much dissatisfaction as an undersized unit.

Chapter 5—Simplified Cooling Calculations—takes you step by step through the procedures necessary for evaluating your home's cooling requirements right down to the completion of the "Residential Cooling Load Form".

Completion of this step will enable your dealer to supply you with the correct-capacity cooling equipment, exact costs, and detailed instructions for installing the equipment in your home. Your kit will look something like the one in Fig. 7-2 and will include two lengths of refrigerant tubing, four baffles, plenum cover, coil supports, sealers, cooling coil, condenser/compressor unit, and complete instructions.

If you still aren't certain you want to tackle this job yourself, briefly scan the following pages. In doing so, you will gain a general knowledge of what the installation procedures

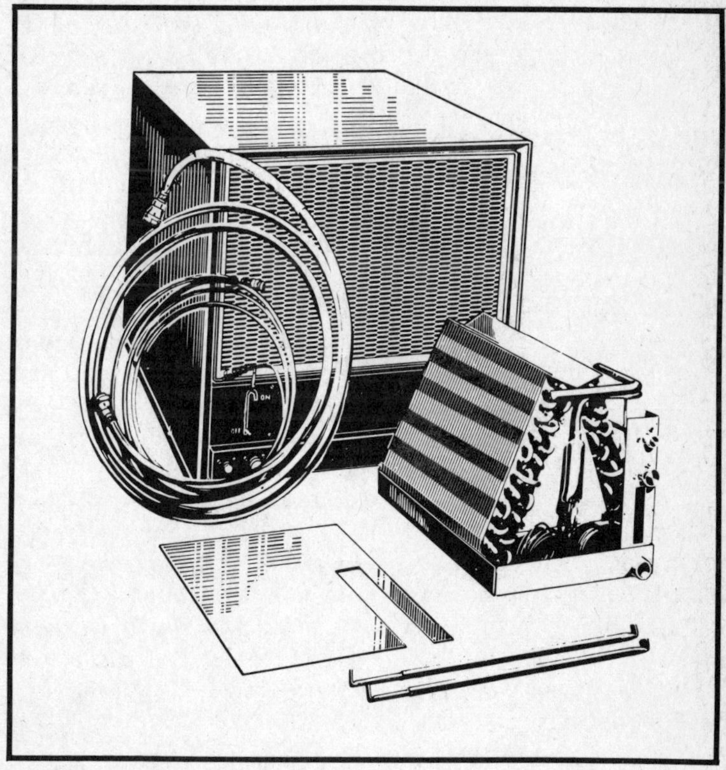

Fig. 7-2. Your kit will look something like this when it arrives.

are. If you decide to do it, a more detailed reading will show you exactly how to go about it.

MOUNTING THE CONDENSING UNIT

Before ordering your kit, select the exact spot for mounting the condensing unit so you will know the length of refrigerant tubing to order.

In general, the condensing unit should be mounted outside the building, as close as possible to the cooling coil inside the furnace plenum, and on solid, level supports such as those shown in Fig. 7-3.

For roof installation, use wooden beams (treated to reduce deterioration) or channel-iron supports. Both should be sized to support the weight of the condensing unit and extended beyond the unit to distribute the load on the roof. Any local iron works can make a support frame for you once you provide them with the dimensions of the unit, or you may make the frame yourself using bolt-together channel iron available at your local hardware store. When installing such a frame on bonded roofs, check for special installation requirements.

For ground level installations, use piers or a concrete slab with footers extended below the frost line, wooden beams treated to reduce deterioration, or a channel-iron frame with a suitable base. These are the most common installations and probably the easiest for the handyman to construct (especially mounting the condensing unit on 4-inch × 4-inch wooden beams).

The condensing unit may also be mounted on a wall of your house, using an angle-iron mounting frame dimensioned to fit the condensing unit and provided with supports for attaching the frame to the building.

If it is necessary to mount the condensing unit more than, say, two feet from the building, you should also make provisions at this time for the electrical and refrigerant tubing connections. An electrical conduit will be required for the power supply and a separate conduit for the low-voltage thermostat wiring. However, if the unit is mounted within two feet of the building, the fused disconnect switch can be mounted on the side of the building and the wiring and refrigerant lines can loop directly from the side of the house to the condensing unit.

When locating the condensing unit, clearance must be provided for the air intake and air discharge, refrigerant piping and power connections, and maintenance and servicing access. The recommended minimum clearances can vary from 6 inches to 6 feet; check the recommendations of the manufacturer.

1. Poured or pre-cast leveled concrete slab, 26" x 26" long x approximately 4" thick.

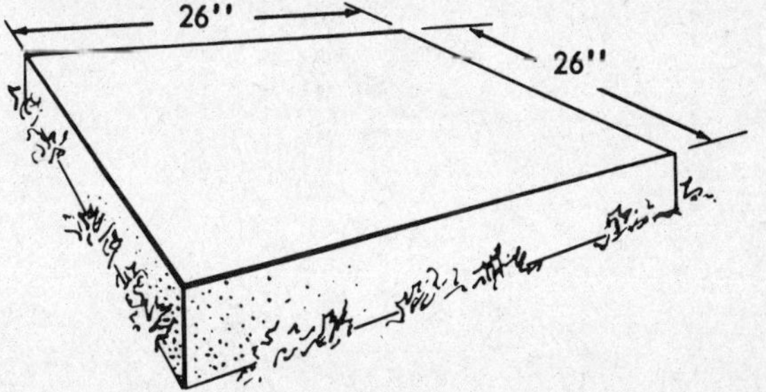

2. Two pre-cast patio blocks approximately 12" x 26" x 4" thick, leveled.

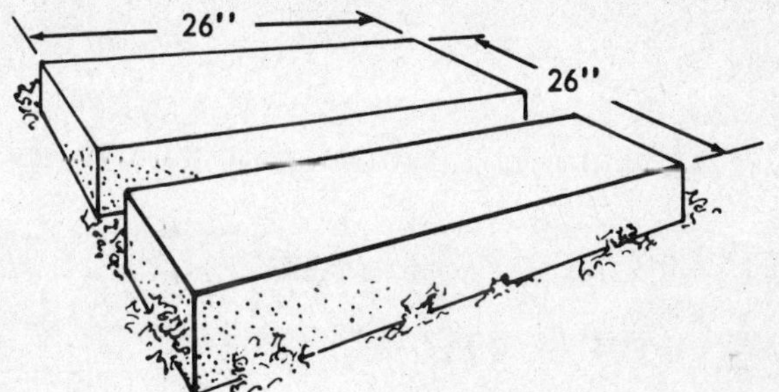

3. Four, 4" x 8" x 16" concrete blocks, set and leveled as shown.

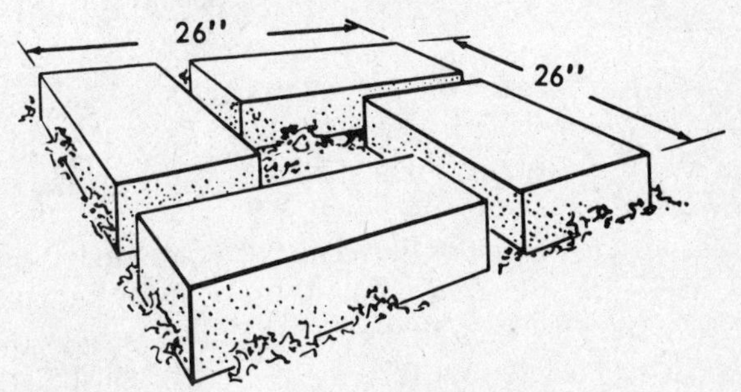

Fig. 7-3. Supports on which the condensing unit is mounted.

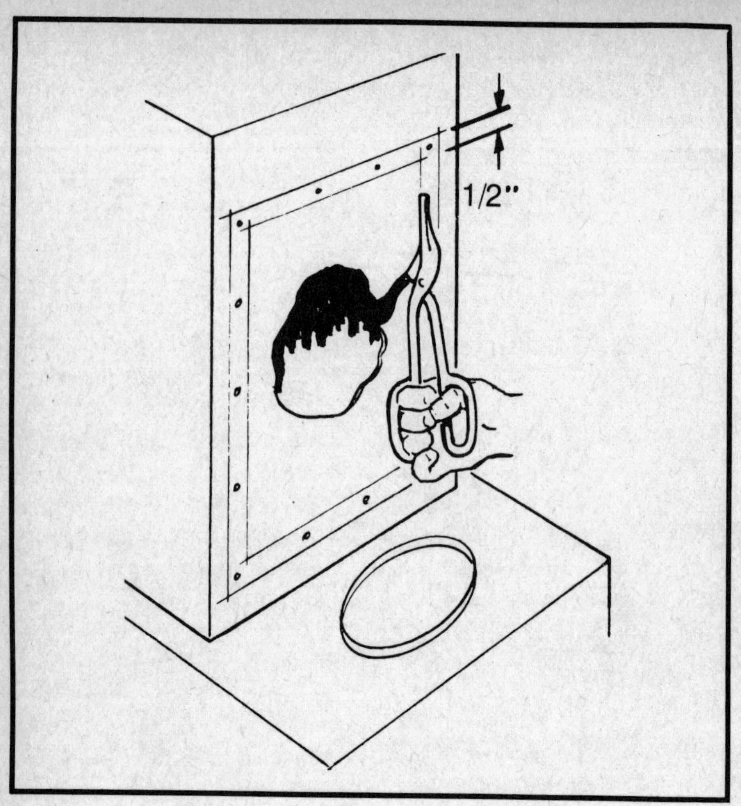

Fig. 7-4. Cutting the opening in the furnace plenum with tin snips.

INSTALLING THE COOLING COIL

Most cooling coils designed for the do-it-yourselfer are charged with a holding charge of refrigerant and sealed at the factory to help simplify the installation. However, always follow the procedure instructions recommended by the manufacturer.

The installer will need only a few hand tools for this job: tin snips, electric drill, screwdriver, knife, hammer, scratch awl, level, drop light, and 1/8-inch drill bits.

Many forced-air heating systems installed within the past 10 or 15 years have provisions in the furnace plenum for adding a cooling system. For instance, one or more of the side panels of the plenum may be secured with sheet-metal screws rather than flanged together. If your furnace plenum is constructed in this manner, the installation of the cooling coil will be even easier. Merely remove the sheet-metal screws and lift the panel off; this will give complete access to the inside of the plenum.

If your furnace does not have this provision, you will have to cut an opening in the plenum. Start the hole by striking a screwdriver or cold chisel at an angle against the plenum. Proceed to cut a hole approximately ten inches in diameter to enable you to see inside the plenum.

All forced-air furnaces contain duct flanges, but they vary in size from furnace to furnace. Therefore, once the opening is made, insert your drop light and accurately measure the height of the flanges on your furnace. Scribe a horizontal line 1/4-inch higher than the flange on the outside of the plenum. Scribe a second level line 1/2-inch below the first.

Line the bottom of the sheet-metal cover that came with your kit up with the second (lower) line just scribed. Center the cover horizontally and mark the location of the holes in the cover on the plenum. Remove the cover and drill the plenum holes with a 1/8-inch drill. Next, hold the cover in position and scribe a line around its perimeter. Remove the cover and scribe a line 1/2-inch inside the perimeter line.

As you will later be cutting along this inside line, make certain that the cooling coil will fit through the resulting hole. When you are certain of the dimensions of the lines, cut along the inside line with tin snips as shown in Fig. 7-4.

Place the coil supports that come with most kits inside the opening with the large-diameter tube first. Hook the flanged end of the supports over the furnace flanges as shown in Fig. 7-5. As the end of the support will be located between the duct flange and the plenum wall, be certain that you keep the support end well down as shown in Fig. 7-5.

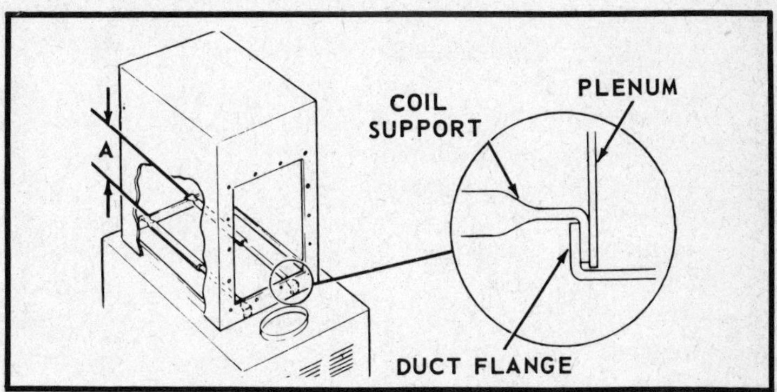

Fig. 7-5. Placement of the coil supports.

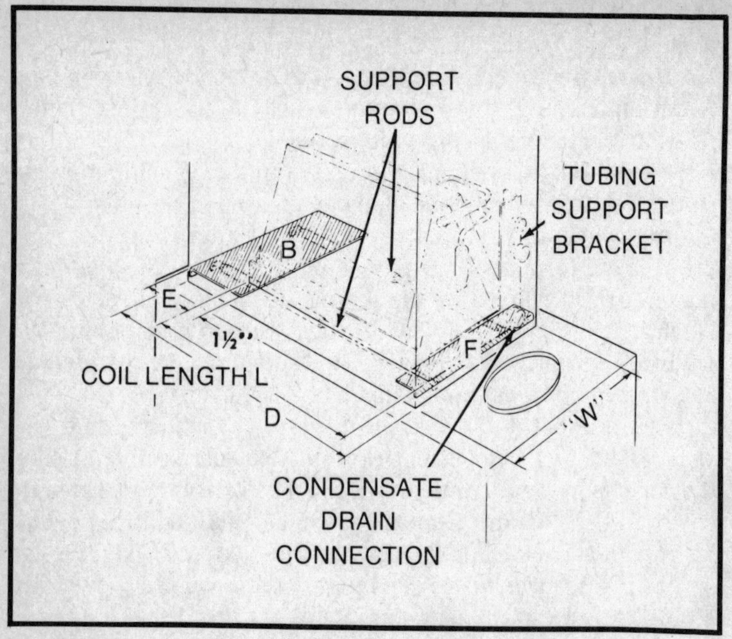

Fig. 7-6. Method of positioning the coil supports inside the plenum.

Some plenums are secured to the duct flange with sheet metal screws, so don't try to locate the support end over a screw. Distance "A" in Fig. 7-5 (between the two supports) must be less than the width of the coil it will support.

With the support ends well-secured between the duct flange and the plenum, telescope the small-diameter portion of the coil support toward the opposite side of the plenum and insert the flanged end over the duct flange.

Refer to Fig. 7-6 to help you visualize the next few steps:

a. Measure depth of plenum (dimension "D").
b. As the front of the tubing support bracket of the coil assembly will always be positioned flush with the front of the plenum, it follows that the coil length ("L") should be subtracted from the plenum dimension ("D") to determine the distance between the back of the coil and the plenum (dimension "E").
c. Select one of the wide baffles supplied with most kits and prepare to cut it to fit shaded area "B." As the baffle should fit under the condensate tray, add 1 1/2-inches to dimension "E" and scribe line No.1 with an awl as shown in Fig. 7 7.

d. Measure the inside width of the plenum, subtract 3/16 of an inch for clearance and scribe line No. 2 . The shaded area in Fig. 7-7 should be the correct size to fit area "B" in Fig. 7-6. Recheck the dimensions to be certain that all measurements are correct, then cut along the scribed lines. Place the baffle in position with the short flange down and against the plenum as shown in Fig. 7-6. If it is impossible to place the short flange down, it may be placed in upright, provided additional care is taken when sealing compound is applied later in the installation.

e. Select the smallest of the remaining baffles. Scribe and cut it in similar fashion to the first baffle to fit in "F" (Fig. 7-6).

f. Measure the width of the air inlet at the base of the cooling coil, and subtract this figure from the width of the plenum. The result divided by two gives you the width of each of the side baffles (L and R in Fig. 7-8). Their length is, of course, the depth of the plenum (dimension D in Fig. 7-6) minus a quarter-inch or so for ease of fit. The purpose of the end and side baffles is to ensure all air moved by the blower goes through—not around—the coil frame.

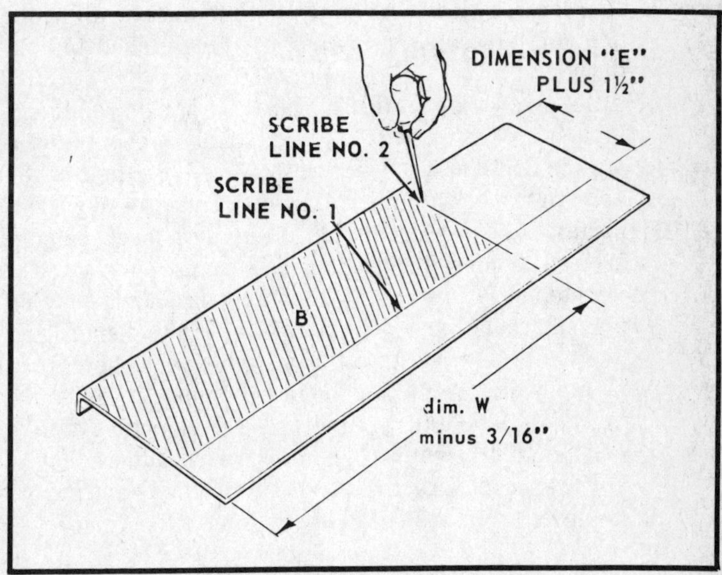

Fig. 7-7. Scribing lines on baffles.

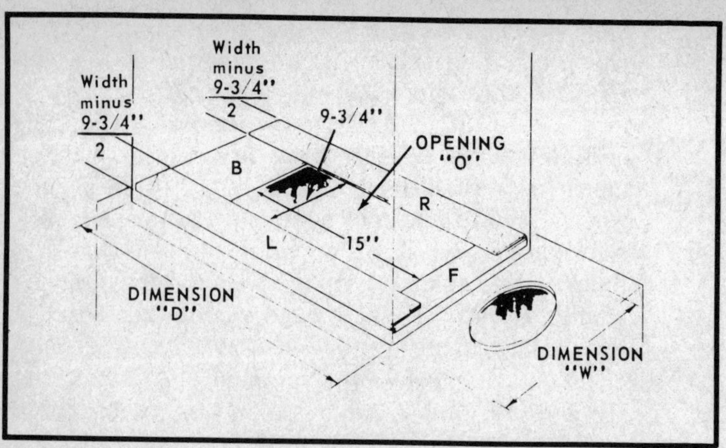

Fig. 7-8. Location and dimensions of side baffles in plenum.

g. As shown in Fig. 7-8, place the baffles inside the plenum with the short flange down and against the sides of the plenum. The baffles should be placed in this sequence: first, B; second, F; third, L; and fourth, R.

h. Opening "O" in Fig. 7-8, framed by the baffles, should not be smaller than the air inlet to the coil. Furthermore, opening "O" should be about an inch smaller in both width and depth than the overall width and depth of coil used, thus providing a shelf to hold the coil.

i. In the event that baffles L and R do not lie relatively flat on baffles B and F, sheet-metal screws may be used to hold them flat. Place the screws so their heads don't interfere with the coil frame. This procedure will normally be required when there are large plenum areas and, consequently, large overlaps.

j. High-temperature caulking compound should now be placed around the entire perimeter of the plenum and worked or pressed into the area between the baffles and plenum. A second bead of caulking should be placed around the perimeter of the opening. Make certain that the caulking compound has adhered to the baffles; to ensure this, a slight pressure should be applied on top of the bead of caulking with a finger.

To place the coil inside the plenum, grasp the coil by the tubing at point A in Fig. 7-9 and the condensate tray at point B.

Always keep the top of the coil near the top of the plenum opening while inserting it; take care that the bottom of the coil doesn't move the bead of caulking out of position.

The front surface of the condensate tray should be flush with the front of the plenum before it is lowered to rest on the caulking strips. After it is gently lowered into place, hold your straight edge or level against the front of the plenum to check it.

Once the coil is in place, try the plenum cover over the opening, matching the holes of the cover with holes in the plenum; the tubing access holes on both sides of the cover should now be over the holes in the tubing support bracket. If they aren't, carefully move the coil until they are lined up. When you move the coil, observe the caulking. If the caulking does not seal between the flanges and condensate tray, you will have to remove the coil and reposition the caulking. A bead of caulking should also be placed between the tubing support bracket and the cooling coil cover to ensure a good seal (see Fig. 7-10). When you are certain that an air-tight seal has been accomplished, replace the cover, insert and tighten all screws.

At this stage, a 3/4-inch condensate line should be connected to the lowest opening and a J-trap should be fabricated to prevent air leakage through the line. The remainder of the pipe should be pitched at least 1/4-inch per 10 feet to an open floor drain, sump pit, or similar place where the condensate water can be disposed.

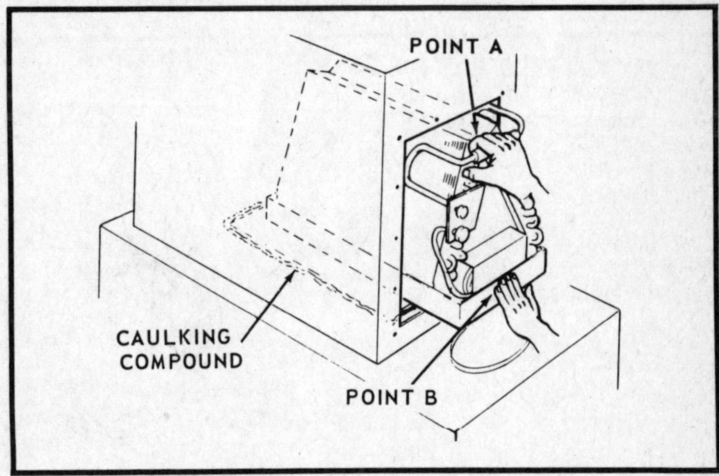

Fig. 7-9. Method of inserting coil into a plenum.

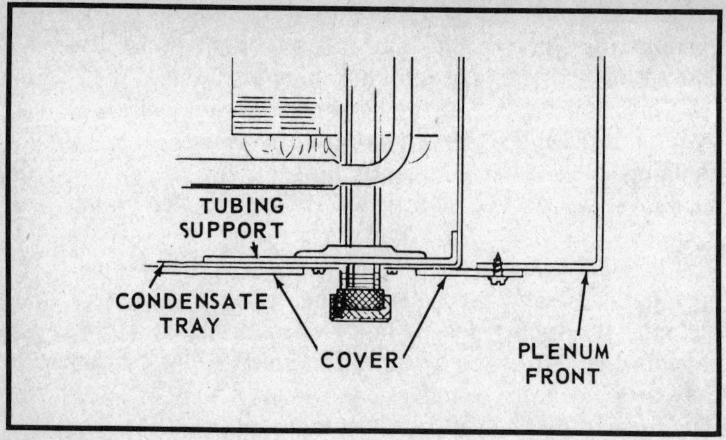

Fig. 7-10. Caulk well between the tubing support bracket and cooling-coil cover.

INSTALLING THE REFRIGERANT TUBING

Most refrigerant tubing that comes in kits for home owners has been specially tempered, evacuated and partially filled with refrigerant at the factory. A typical fit will consist of:

1. One coil of small copper tubing with couplings.
2. One coil of large copper tubing with spring tube bender, insulation and couplings.

Before attempting to install the refrigerant tubing, inspect it carefully, especially where the couplings are brazed to the

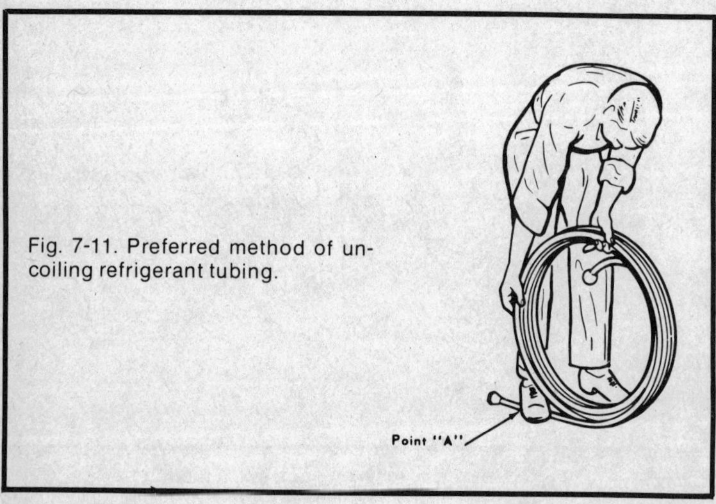

Fig. 7-11. Preferred method of uncoiling refrigerant tubing.

ends. If any openings or punctures are detected, have them repaired or replaced.

Once the path of the tubing between the cooling coil and the condensing unit has been determined, the coils may be straightened. A preferred method of uncoiling the tubing is to hold the coil upright, placing one of the ends on a smooth surface (see Fig. 7-11). Hold one end at point A, and proceed to unroll the coil as shown.

Since the cooling coil is usually inside the house and the condensing unit outside, two 2-inch holes must be cut in the wall for the refrigerant lines.

You will have to do some bending of the tubing, but since the copper tubing has been annealed, it can be easily and neatly formed. The smaller tubing can be formed by hand, the bends made over the thumbs as shown in Fig. 7-12. None of the bends, however, should have a radius smaller than 6 inches.

For the larger tubing, use the tubing bender; this bender is installed on the straight end of the tubing and under any insulation. The bend, with the spring bender covering the area of the bend, should then be made carefully over the knee as shown in Fig. 7-13. In making bends, extreme care must be taken to avoid kinking the tubing, which will make it useless.

Remove the protector plugs from the tubing couplings and the protector caps from the couplings of the condensing unit. Make sure that the rubber seal is inside the male coupling of the outdoor unit; if not, immediately obtain one from your supplier as it *must* be used.

Without using wrenches, thread the couplings of the tubing over the respective couplings of the condensing unit. Tighten finger tight. The tubing should now be formed and routed to the

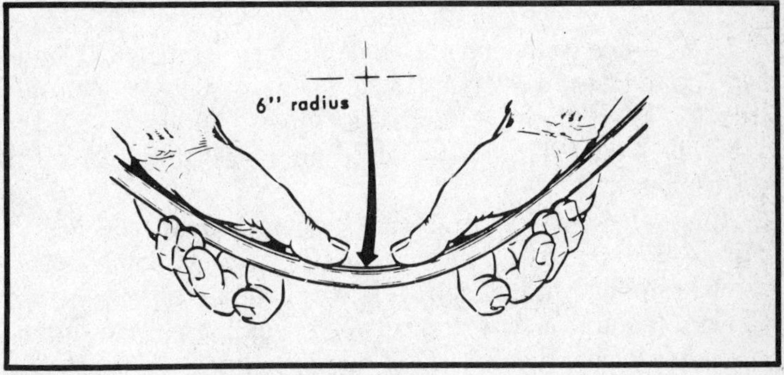

Fig. 7-12. Method of bending the smaller tubing.

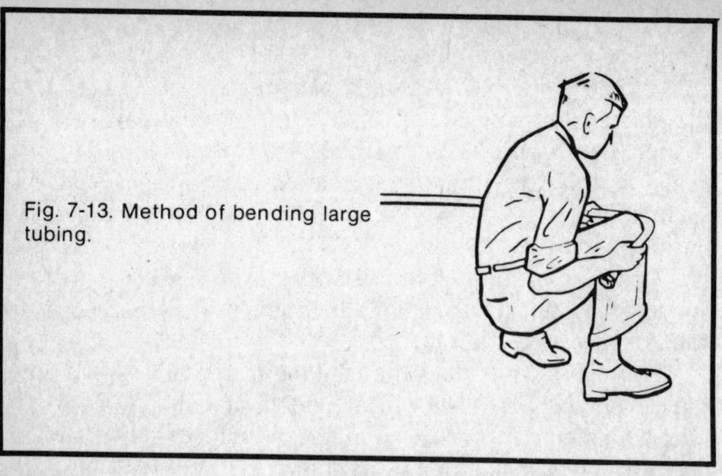
Fig. 7-13. Method of bending large tubing.

cooling coil inside the house. It is recommended that the larger tubing be formed and run first, and the small tubing then run adjacent to it.

Next remove the protective caps from the cooling coil; again make certain that the rubber seal is in palce within the fittings. Remove the protector plugs from the female tubing connection and connect with finger pressure—no wrenches.

The exact method of connecting the refrigerant tubing to the units will vary from unit to unit, but the following example will illustrate the basic process.

Using appropriate wrenches as shown in Fig. 7-14, tighten the couplings in this sequence:

1. Large tube to inside cooling coil.
2. Small tube to inside cooling coil.
3. Large tube to outdoor unit.
4. Small tube to outdoor unit.

Tighten the couplings until a definite resistance is felt and then tighten an additional 1/6 to 1/4 of a turn. If a torque wrench is used, the small-diameter tube should be torqued to between 10 and 12 foot-pounds; the large tube, 35 to 45 foot-pounds.

Immediately after tightening each coupling, coat the entire coupling and the brazed joint where the tubing connects to the coupling with a rich mixture of liquid detergent and water. If leakage is detected between coupling halves, tighten the coupling nut. If this doesn't work, contact your dealer or supplier at once.

INSTALLING THE ELECTRIC WIRING

Follow instructions given in Chapter 4 for installing the electric power circuit. Remember that all work must be done in accordance with the National Electrical Code.

Again, different units will have slightly different wiring connections; the wiring diagram in Fig. 7-15 shows a typical hookup.

If there is any doubt at all about your capability in this phase of the installation, call in an electrician to do the job for you. It should cost around $100 for the wiring, provided your electric service entrance is adequate for the additional load.

PRESTARTING CHECK LIST

Before starting the unit, check the following details in order to prevent possible damage.

1. Is there adequate clearance around the condensing unit and its related grilles?
2. Is the electrical wiring in accordance with the wiring diagrams furnished with the system?
3. Are all wiring connections tight?
4. Does the blower turn freely without striking the shroud?
5. Is the thermostat level and in a proper location, that is, away from drafts and on an inside wall or partition?

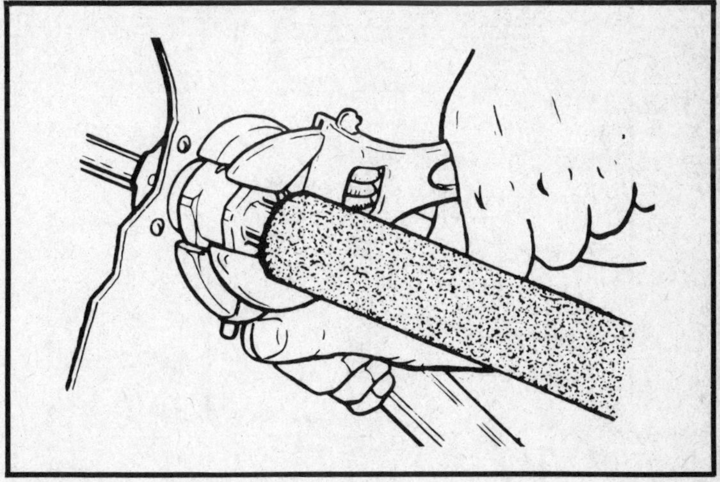

Fig. 7-14. Correct use of wrenches to tighten couplings on refrigerant lines.

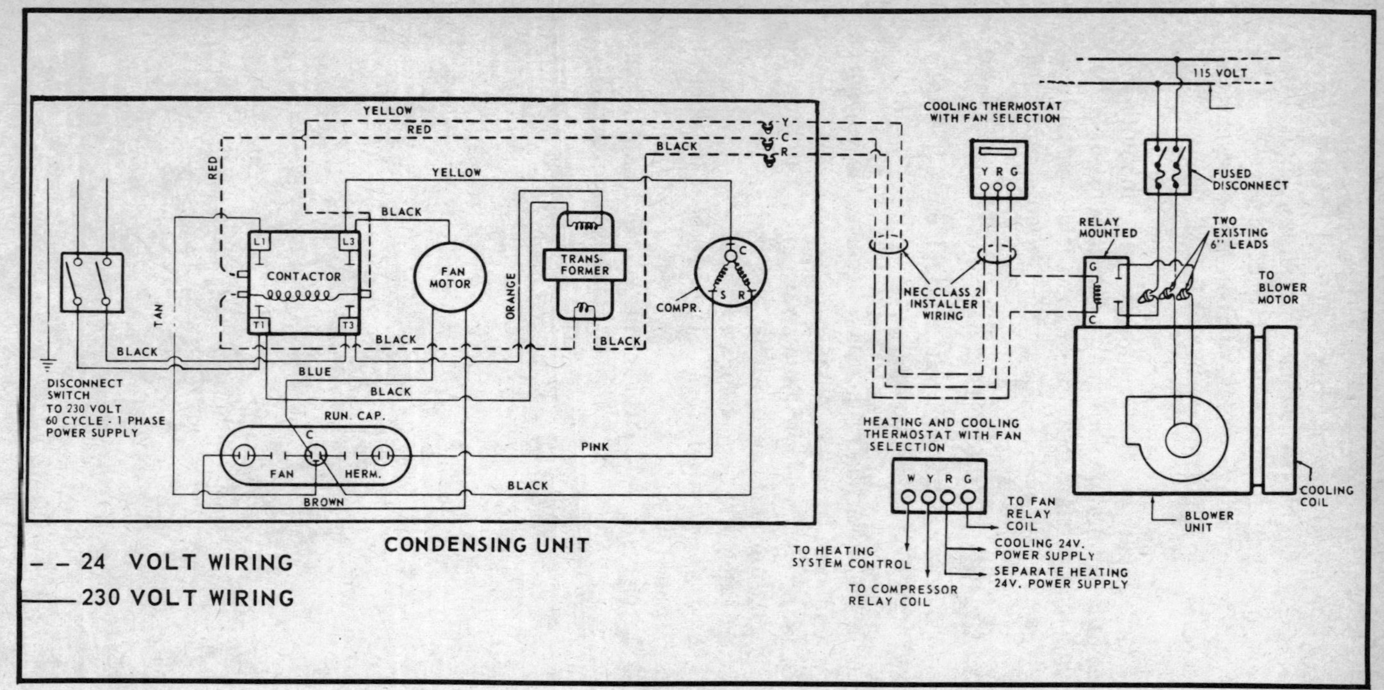

Fig. 7-15. Electrical wiring diagram of a typical air-conditioning system.

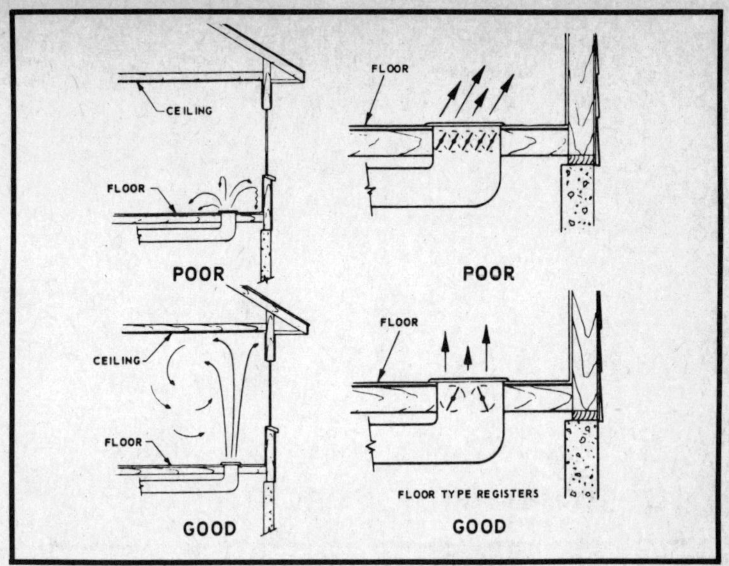

Fig. 7-16. All register dampers should be adjusted so that the air projects upward.

6. Are the filters or filter of adequate size, clean and in place?
7. Is the ductwork correctly installed, well-insulated and taped?

START-UP PROCEDURE

Here's the moment you've been waiting for. To start, first turn the thermostat to its highest setting, and turn on the power at the electric panel or safety switch. Then go from room to room and adjust all registers or diffusers so that the air projects upward as shown in Fig. 7-16. Then move the thermostat indicator to a setting below room temperature, at which time the blower and refrigerant system should operate.

The system may need balancing to provide the greatest volume of air in areas with the greatest exposure; this is accomplished by a trial-and-error method, using the dampers inside the ductwork for balancing.

Thoroughly read any literature that comes with the unit before and after installation. Most come with booklets concerning troubleshooting and maintenance of the equipment, so they're worth holding on to.

You can see that what was once a job for professionals is now so simple you can do it in a single day.

Chapter 8

Installing Your Own Ductwork

Many existing homes without air conditioning are heated with steam or hot water and use pipes rather than ductwork for the circulation of heat. While it is somewhat more difficult (and expensive) to install air conditioning in these homes than in homes with ductwork, there are still several practical ways to get the job done.

You may install thru-wall heating and cooling units as described in Chapter 6, using your existing piping. Or you could install a self-contained air-conditioning unit as shown in Figs. 8-1 and 8-2 with a new duct system, entirely independent from your heating system. When many people think of ductwork, they have visions of spot welders, sheet metal brakes, and other expensive tools running into thousands of dollars. This is not the case with fiberglass ducts: they're within the capabilities of nearly every homeowner!

Unlike ductwork constructed from sheet metal, fiberglass duct requires only a few common hand tools for assembly and installation. They include a sharp knife, a hammer, pliers, a heat-seal tool, butt-joint tools, and a staple gun. If you have to purchase all of them, they should run you less than $35.

Furthermore, a conventional sheet-metal system requires consideration of thermal efficiency, acoustical efficiency and vapor resistance as well as air delivery. All these features must be engineered and installed in a step-by-step process—enough to discourage any do-it-yourselfer from the

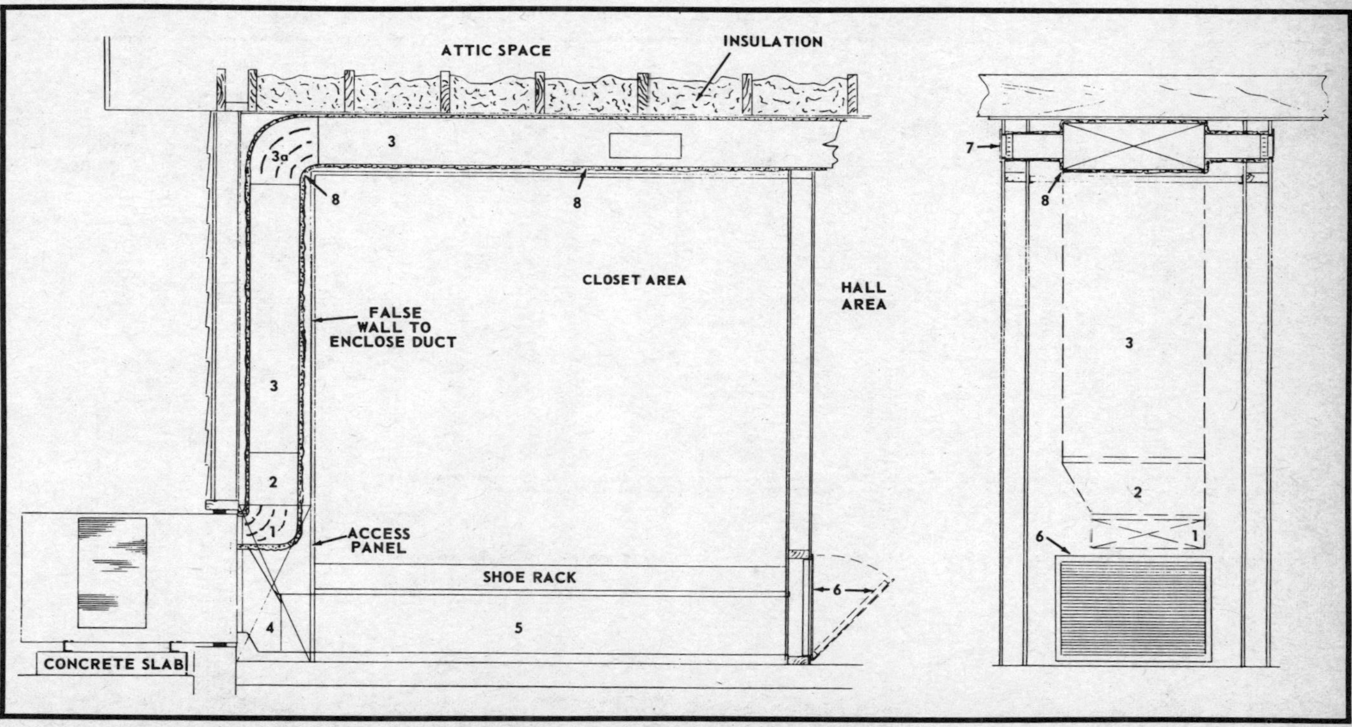

Fig. 8-1. Combination heating and cooling unit mounted on concrete slab, projecting through the house wall.

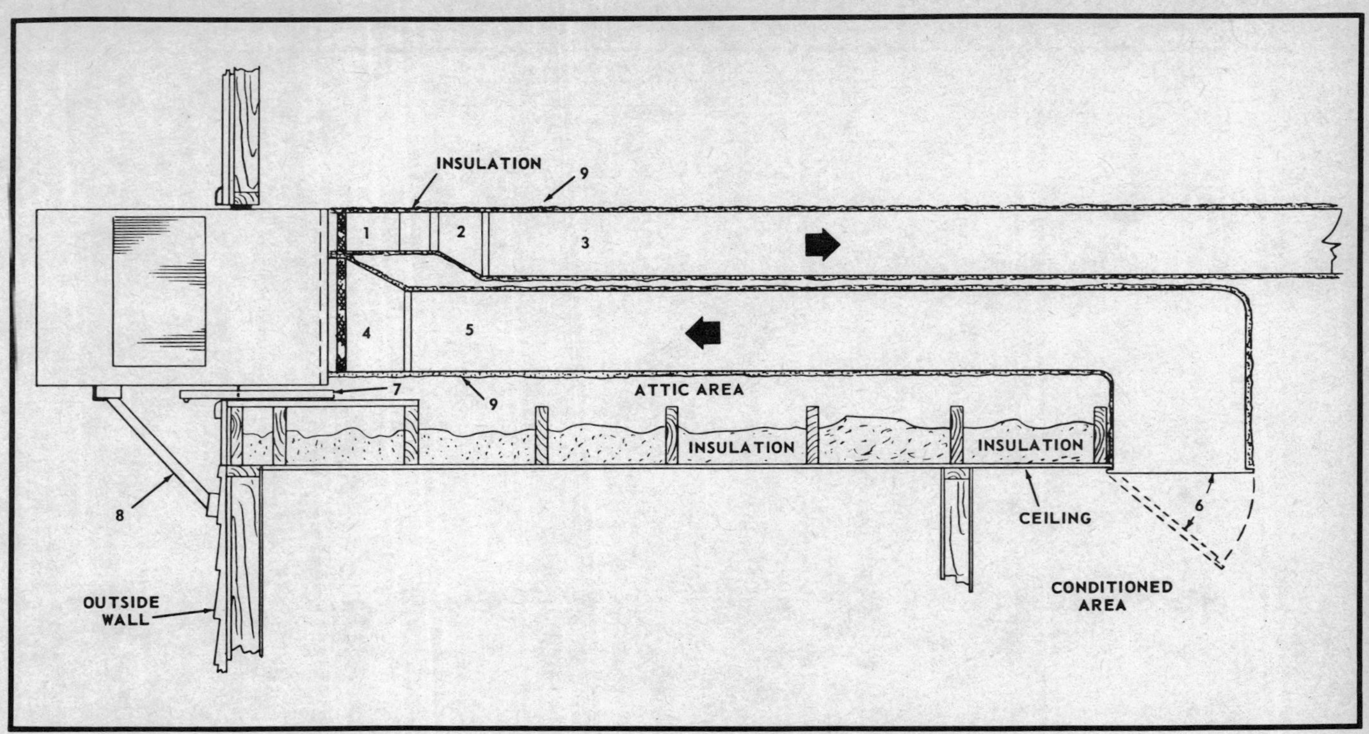

Fig. 8-2. Combination heating and cooling unit mounted through the wall in attic area.

start. Not so with fiberglass duct! It's an air conveyor which has insulation, sound absorber, and vapor barrier built in. By incorporating these custom-engineered system requirements into a single product, fiberglass ductwork not only reduces the time, money, and effort required to design and install ducts, but also assures that the system will have minimum loss, noise, and condensation.

ROUND DUCT

Prefabricated round duct is quickly and easily installed with standard sheet metal fittings, a knife, heat-seal iron, factory-supplied templates, and sealing tape. The sheet metal fittings assure proper alignment and provide reinforcement at the joints, which are also secured with pressure-sensitive aluminum tape and heat sealed. This assures continuation of the vapor barrier and strong, air-tight joints. Should a puncture occur in the vapor barrier, it can easily be mended by applying tape over the damaged area.

RECTANGULAR DUCT

Rectangular fiberglass duct is easy to use: even the largest sections can be positioned easily due to their light weight and resilience. Sheet metal connectors are not required for joining rectangular duct sections unless the duct span exceeds the maximum allowable span for a specific thickness of duct board at a given pressure. However, rectangular duct runs do require suspension on 6-foot (maximum) centers unless the duct can be supported by structural members.

COMBINED DUCT INSTALLATIONS

Round duct take-offs from rectangular duct are easily accomplished with fiberglass duct. A round sheet-metal fitting the same diameter as the round duct take-off is secured to the rectangular trunk. The round take-off is then slipped over the metal fitting and pushed tightly against the rectangular trunk, taped around the curvature of the connection, and heat-sealed. It's just as simple as that.

PLANNING THE INSTALLATION

Begin by making a careful analysis of your home's present and future cooling requirements. (Chapter 5 of this book takes you step by step through this procedure, right down to the

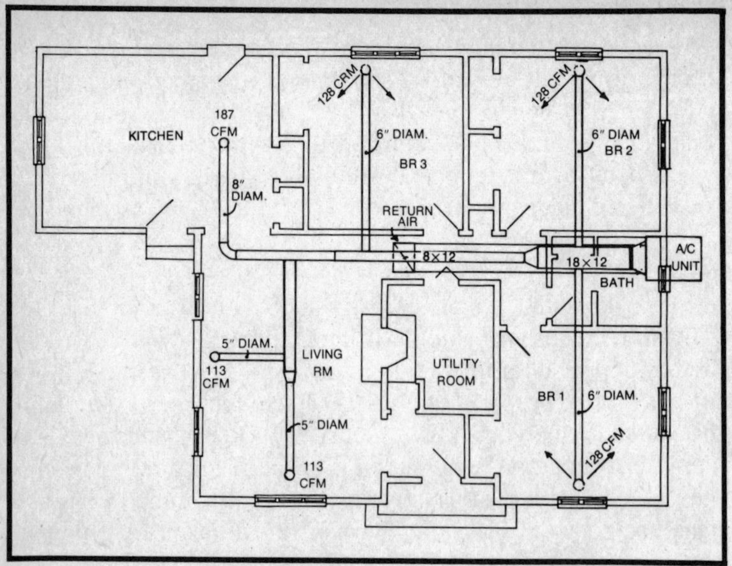

Fig. 8-3. Floor plan of residence with ductwork laid out.

completed "Residential Cooling Load Estimate Form.") This step shows what size unit you will need, and also helps to determine the size of ducts needed to serve various areas.

It would take more space than we have to show how to size the ductwork for your cooling system. However, if you follow the suggestions given here, you will be able to have an engineer size the system for you at little or no cost. Try to find a floor plan of your home, but if this is not possible, take measurements and sketch one of your own. Include all windows, doors, lighting fixtures, large beams, and any other obstruction that might interfere with the running of the ductwork. With the sketch and a copy of your cooling calculations, visit your local utility company and ask if one of the engineers could size and lay out the ductwork for you. Since you will be buying more current from them because of the new cooling system, they will probably be happy to help you.

Another possible advisor is the manufacturer of the cooling unit you intend to purchase or his factory representative. Make copies of your sketch and cooling calculations and mail them to the manufacturer. Be sure to indicate the exact location of the self-contained cooling unit; that is, on the ground (Fig. 8-1), through the wall (Fig. 8-2), etc. Tell them you are planning to add an air-conditioning system to your home—utilizing their equipment—and would appreciate their laying out the ductwork for you.

Regardless of who designs the duct system,the final layout should appear something like Fig. 8-3. This drawing includes all windows, doors, wall partitions, the location of the self-contained cooling unit, the routing of ductwork, the location of grilles and diffusers, and the size of each. The drawing should be drawn to scale, and it might even include a section through your house to help you visualize the system better.

When dimensions appear on rectangular duct, say, 24 inches × 16 inches, the first dimension always means the width of the duct, and the second dimension, the depth of the duct. Dimensions given for round duct, like 8 inches θ means the inside diameter of the duct.

Select supply-air outlets carefully; they are a critical part of any air distribution system. The ideal supply-air outlet will deflect or diffuse the air silently, adjust to change the air-flow rate, and be able to throw the conditioned air no less than three-quarters of the distance from the outlet to the opposite wall. Outlets are located according to the shape, size, and heat load of the area.

Air-conditioning systems also require grilles and ducts to return air to the cooling unit. While the location of a return is not as critical as that of a supply output, it should be located on the side of the room opposite to the supply. With this information, and using Fig. 8-3 as a guide, we can begin ordering the materials and installing the system.

We will assume that the self-contained air-conditioning unit has been installed in the attic of the home according to the manufacturer's instructions, wiring has been performed in accordance with the National Electrical Code, and everything is ready for the ductwork.

FABRICATING FIBERGLASS DUCT

The section of ductwork closest to the unit (Fig. 8-3) is rectangular—18 inches wide by 12 inches deep—and will be fabricated from a flat fiberglass duct board. First determine the width of the board required. Board width is equal to twice the interior duct width (2 × 18 or 36 inches in our case) plus twice the interior height (2 × 12 or 24 inches in our case), plus an allowance for grooving. The table in Fig. 8-4 may be used for quick selection of

THICKNESS		HEIGHT																						
1½"			6	7	8	9	10	11	12	13	14	15	16	17	18	19	20	21	22	23	24	25	26	27
	1"	6	7	8	9	10	11	12	13	14	15	16	17	18	19	20	21	22	23	24	25	26	27	28
WIDTH																								
	6	32	34	36	38	40	42	44	46	48	50	52	54	56	58	60	62	64	66	68	70	72	74	76
6	7	34	36	38	40	42	44	46	48	50	52	54	56	58	60	62	64	66	68	70	72	74	76	78
7	8	36	38	40	42	44	46	48	50	52	54	56	58	60	62	64	66	68	70	72	74	76	78	80
8	9	38	40	42	44	46	48	50	52	54	56	58	60	62	64	66	68	70	72	74	76	78	80	82
9	10	40	42	44	46	48	50	52	54	56	58	60	62	64	66	68	70	72	74	76	78	80	82	84
10	11	42	44	46	48	50	52	54	56	58	60	62	64	66	68	70	72	74	76	78	80	82	84	86
11	12	44	46	48	50	52	54	56	58	60	62	64	66	68	70	72	74	76	78	80	82	84	86	88
12	13	46	48	50	52	54	56	58	60	62	64	66	68	70	72	74	76	78	80	82	84	86	88	90
13	14	48	50	52	54	56	58	60	62	64	66	68	70	72	74	76	78	80	82	84	86	88	90	92
14	15	50	52	54	56	58	60	62	64	66	68	70	72	74	76	78	80	82	84	86	88	90	92	94
15	16	52	54	56	58	60	62	64	66	68	70	72	74	76	78	80	82	84	86	88	90	92	94	96
16	17	54	56	58	60	62	64	66	68	70	72	74	76	78	80	82	84	86	88	90	92	94	96	98
17	18	56	58	60	62	64	66	68	70	72	74	76	78	80	82	84	86	88	90	92	94	96	98	100
18	19	58	60	62	64	66	68	70	72	74	76	78	80	82	84	86	88	90	92	94	96	98	100	102
19	20	60	62	64	66	68	70	72	74	76	78	80	82	84	86	88	90	92	94	96	98	100	102	104
20	21	62	64	66	68	70	72	74	76	78	80	82	84	86	88	90	92	94	96	98	100	102	104	106
21	22	64	66	68	70	72	74	76	78	80	82	84	86	88	90	92	94	96	98	100	102	104	106	103
22	23	66	68	70	72	74	76	78	80	82	84	86	88	90	92	94	96	98	100	102	104	106	108	110
23	24	68	70	72	74	76	78	80	82	84	86	88	90	92	94	96	98	100	102	104	106	108	110	112
24	25	70	72	74	76	78	80	82	84	86	88	90	92	94	96	98	100	102	104	106	103	110	112	114

25	26	72	74	76	78	80	82	84	86	88	90	92	94	96	98	100	102	104	106	108	110	112	114	116
26	27	74	76	78	80	82	84	86	88	90	92	94	96	98	100	102	104	106	108	110	112	114	116	118
27	28	76	78	80	82	84	86	88	90	92	94	96	98	100	102	104	106	108	110	112	114	116	118	120
28	29	78	80	82	84	86	88	90	92	94	96	98	100	102	104	106	108	110	112	114	116	118	120	
29	30	80	82	84	86	88	90	92	94	96	98	100	102	104	106	108	110	112	114	116	118	120		
30	31	82	84	86	88	90	92	94	96	98	100	102	104	106	108	110	112	114	116	118	120			
31	32	84	86	88	90	92	94	96	98	100	102	104	106	108	110	112	114	116	118	120				
32	33	86	88	90	92	94	96	98	100	102	104	106	108	110	112	114	116	118	120					
33	34	88	90	92	94	96	98	100	102	104	106	108	110	112	114	116	118	120						
34	35	90	92	94	96	98	100	102	104	106	108	110	112	114	116	118	120							
35	36	92	94	96	98	100	102	104	106	108	110	112	114	116	118	120								
36	37	94	96	98	100	102	104	106	108	110	112	114	116	118	120									
37	38	96	98	100	102	104	106	108	110	112	114	116	118	120										
38	39	98	100	102	104	106	108	110	112	114	116	118	120											
39	40	100	102	104	106	108	110	112	114	116	118	120												
40	41	102	104	106	108	110	112	114	116	118	120													
41	42	104	106	108	110	112	114	116	118	120														
42	43	106	108	110	112	114	116	118	120															
43	44	108	110	112	114	116	118	120																
44	45	110	112	114	116	118	120																	
45	46	112	114	116	118	120																		
46	47	114	116	118	120																			
47	48	116	118	120																				
48	49	118	120																					
49	50	120																						

Fig. 8-4. Board Width Selection Chart.

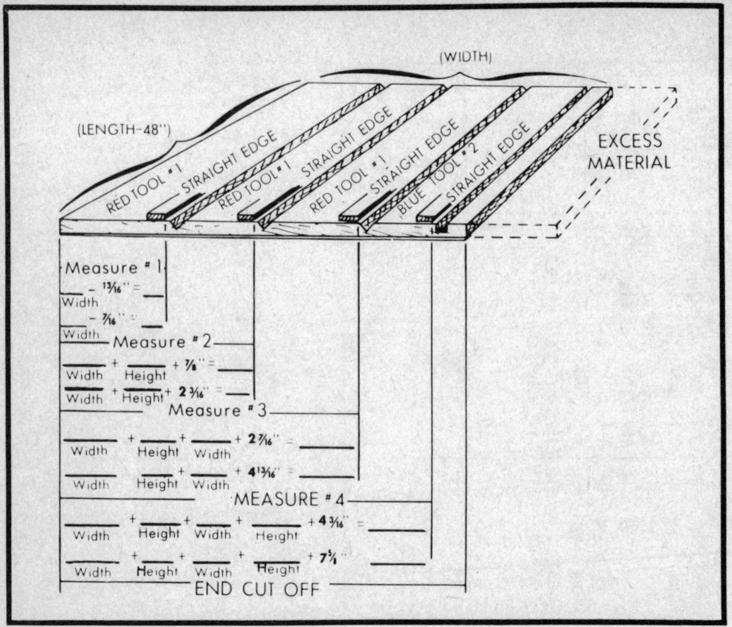

Fig. 8-5. Method and measurements for grooving fiberglass duct board.

board width. The figures have been precalculated and include the amount of material needed for the duct section plus groove allowances.

The left-hand column lists inside duct width in inches, and the figures across the top, inside duct height. Figures are given for both 1-inch and 1 1/2-inch board thicknesses.

To use the chart, find the duct width in the proper thickness column on the left. Then find the duct height in the proper thickness column across the top. Read across and down to the point of intersection, which is the board width necessary.

To illustrate the use of this chart for the 18-inch-by-12-inch duct, locate the 18-inch width figure in the 1-inch (board thickness) column on the left. Locate the 12-inch height figure in the 1-inch height row at the top of the chart. Then read across from 18″ and down from 12 to the point of intersection. The board width is 68 inches. Ducts up to ten feet in length can be fabricated from flat board, providing the total width of the board required does not exceed 70 inches. We will need a fiberglass board 68 inches wide by 10 feet long.

Begin fabrication by marking the board with four measurements as shown in Fig. 8-5. Measure from left to right, using the formula provided on the work sheet in Fig. 8-5; the

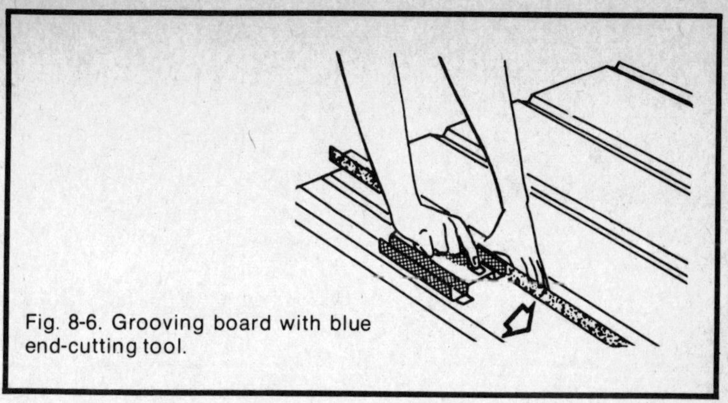
Fig. 8-6. Grooving board with blue end-cutting tool.

three "V" grooves are all made with a red grooving tool (for this brand of fiberglass board). Place the tool with its left edge at each mark and cut along a straight line using a straight-edge as a guide. The groove will appear off to the right of the mark rather than on the mark. Push, don't pull, the tool firmly and evenly to prevent the tool from riding out of the groove or gouging.

When the three "V" grooves have been cut with the red tool, cut the fourth and final groove in the board with the blue tool, pushing in the direction of the arrow in Fig. 8-6, along the mark indicated in Fig. 8-5. The tool will not completely cut through both insulation and jacket.

If a board exactly the right width has been selected, there will be no excess board to remove. Using a utility knife, cut the insulation material from the aluminum jacket by sliding the knife the length of the board between the jacket and the insulation as shown in Fig. 8-7. Cut back only to the first end cutoff groove (approximately two inches).

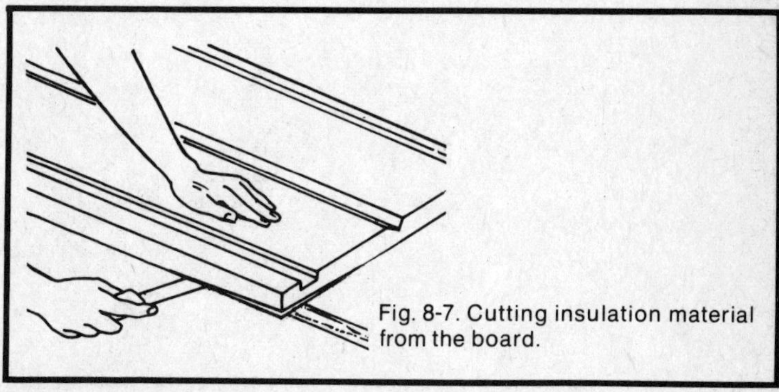
Fig. 8-7. Cutting insulation material from the board.

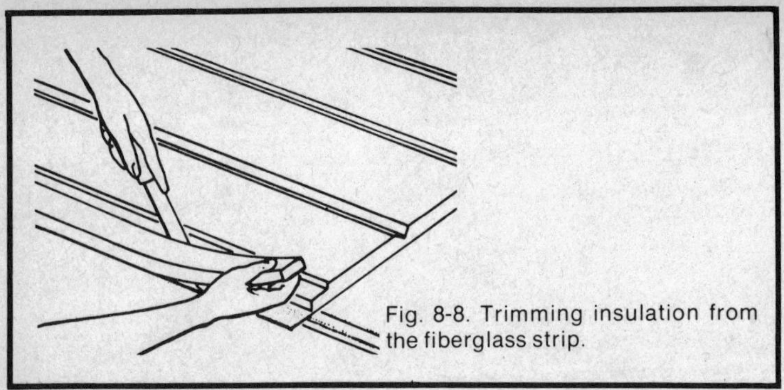
Fig. 8-8. Trimming insulation from the fiberglass strip.

Since the blue tool does not cut completely through the insulation, it is necessary to separate the insulation strip from the side of the last groove with a utility knife—be careful not to cut through the jacket. After removing this excess insulation strip, scrape the adhesive side of the aluminum tab clean with the blade of the knife. (See Fig. 8-8).

Another piece of duct will connect to this first section, so some provision for connecting them is necessary. We'll use a shiplap connection, made as follows:

1. Use the black butt-joint tool and cut along the edge of the board perpendicular to the grooving cuts. The tool must be pushed, rather than pulled, along the edge of the board to insure a correct cut as shown in Fig. 8-9.
2. Next prepare the male shiplap by turning the board over; this is formed on the opposite end of the board from the female shiplap described previously. With the aluminum jacket facing up, insert the utility knife between jacket and insulation and separate the

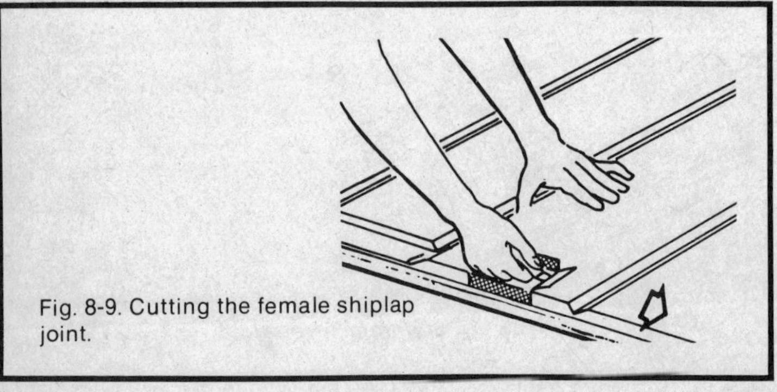
Fig. 8-9. Cutting the female shiplap joint.

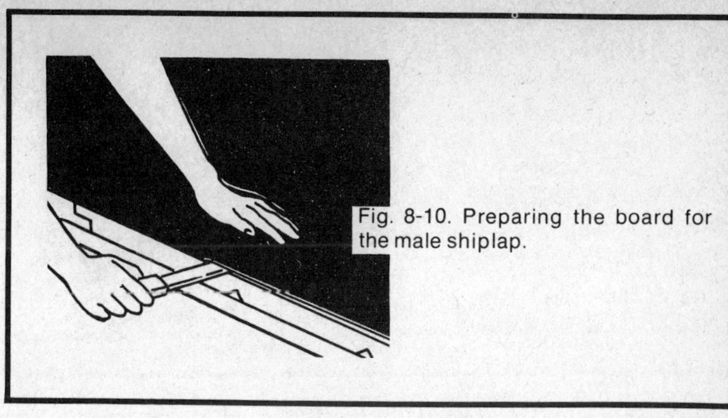
Fig. 8-10. Preparing the board for the male shiplap.

insulation from the jacket for five to six inches from the end of the board. See Fig. 8-10.

3. Again using the black butt-joint tool, make the shiplap cut. Hold the aluminum jacket back far enough to allow freedom of movement in pushing the tool along the edge of the board as shown in Fig. 8-11.
4. Remove the thin remaining tab of insulation from the left edge of the board to insure a snug fit when the board is formed into shape (see Fig. 8-12). The material will tear easily with your fingers. Allow the aluminum jacket to bend back into normal position along the edge of the board.

Now you're ready to make the duct board look like a piece of rectangular duct. Hold the duct so that the overlapping staple tab will fold over the top section as indicated in Fig. 8-13, and draw it into shape using one of the sections as the base of the duct. Now hold this folded duct section against your body

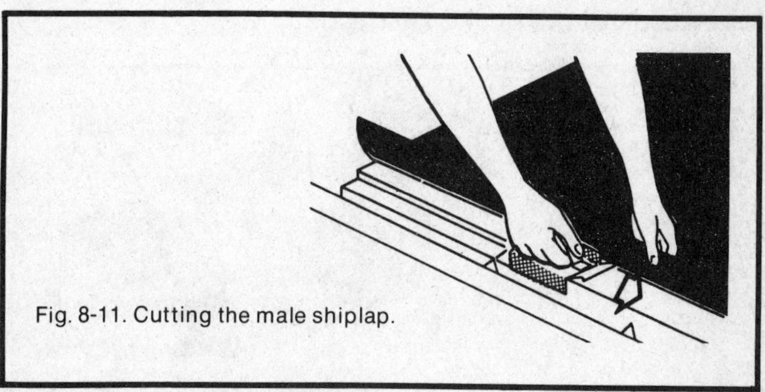
Fig. 8-11. Cutting the male shiplap.

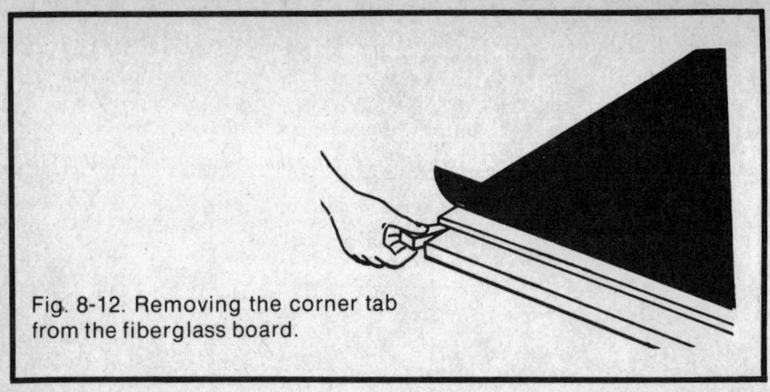

Fig. 8-12. Removing the corner tab from the fiberglass board.

and draw the edges tightly together with the staple tab overlapping the top section. Beginning from the center of the duct, staple this tab (with staples two inches apart and one inch from the edge of the tab) along the full length of the duct

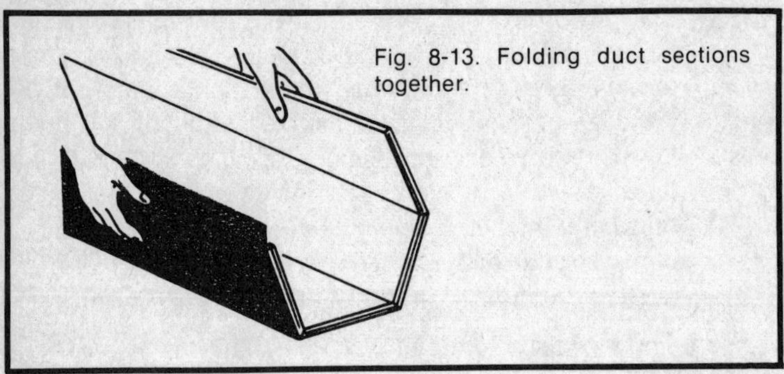

Fig. 8-13. Folding duct sections together.

section as shown in Fig. 8-14. Apply aluminum tape to the joint (Fig. 8-15), making certain that both staples and tab are thoroughly covered. Rub the tape down smoothly along the entire joint to form a neat, tight seal.

Fig. 8-14. Stapling the connecting tab of duct.

The first section of the rectangular ductwork is now complete and may be put in place. If all duct sections were the same size as the first section, it would be necessary only to repeat the procedure given for this first section and tape the sections together. However, in the layout in Fig. 8-3, you will notice that the second section is only 8 inches wide instead of 18 inches. The height is the same—12 inches.

Whenever it becomes necessary to decrease either the height or width of a size of duct, a two-piece transition is constructed. A transition can decrease *either* the height or width of a duct section, but not both. When it does become necessary to decrease both height and width, two separate transitions are utilized.

Using the formulas given in Fig. 8-16 and a tape measure or rule, mark the board with three measurements as shown, measuring from the left edge of the board. Place the correctly-colored tool with its left edge at each mark and cut along a straightedge. Again, the groove will appear to the right of the measurement mark, rather than at the mark. Push, don't pull, the tool firmly and evenly to prevent it from riding out of the groove or gouging the material. The cut at the third measurement is omitted when the board is the exact size required.

After grooving the board and removing excess, if any, prepare the small end of the transition by calculating trim dimensions for the opposite end of the board, measuring in from each edge of the board the required distance. Using a utility knife, cut through the insulation and jacket to remove trim as directed. When the total of measurements A, B, and C exceeds 48 inches, either distance A or C may be varied to reach a total of 48 inches.

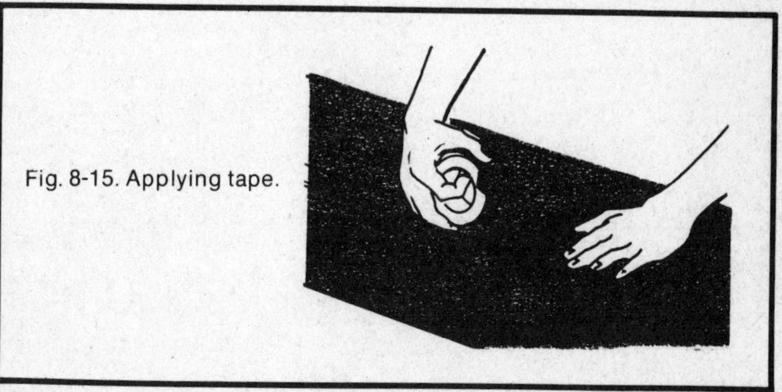

Fig. 8-15. Applying tape.

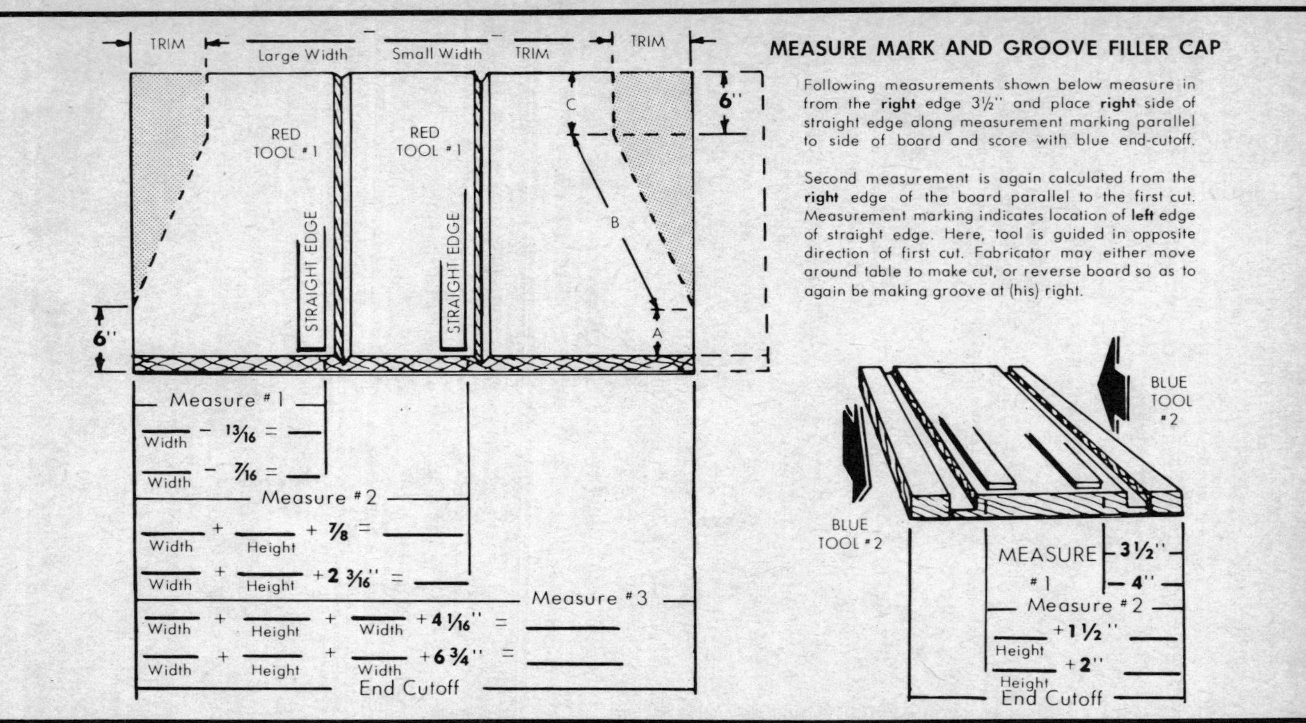

Fig. 8-16. Method of laying out a transition duct.

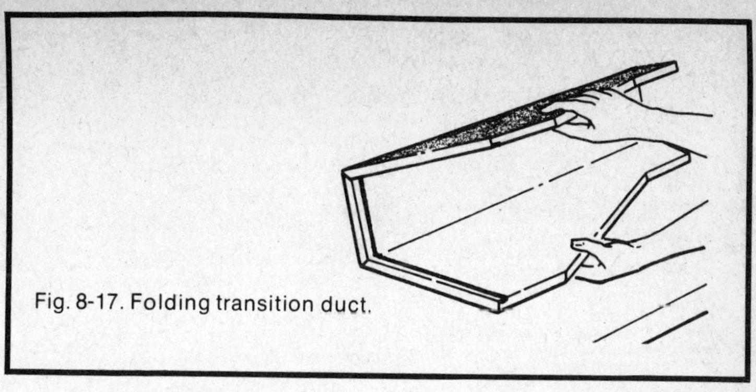
Fig. 8-17. Folding transition duct.

Once the proper cuts have been made, fold the section of the transition with the opening toward your body and the leading edge flush with the work table or floor. Then, starting at the large end, place the filler cap in position so the grooves

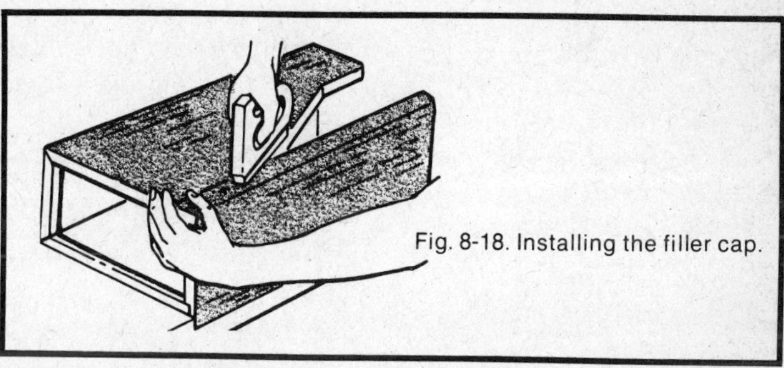
Fig. 8-18. Installing the filler cap.

are aligned with the duct section. Fold the staple tab down flush and begin stapling. Then cut the stapling tab where the filler cap angles in and out to prevent wrinkling. Complete the stapling procedure on both sides, apply tape to the joints and

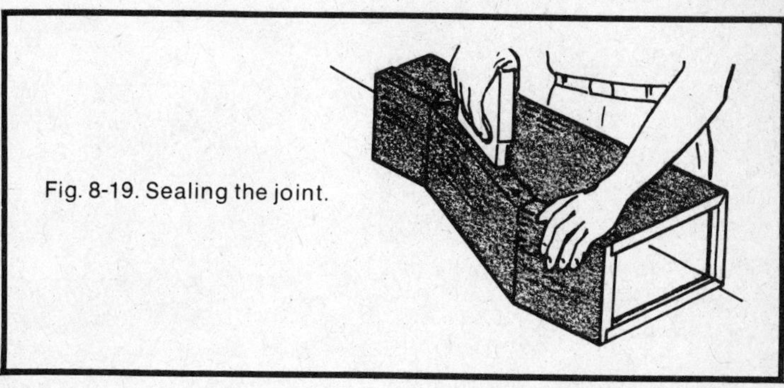
Fig. 8-19. Sealing the joint.

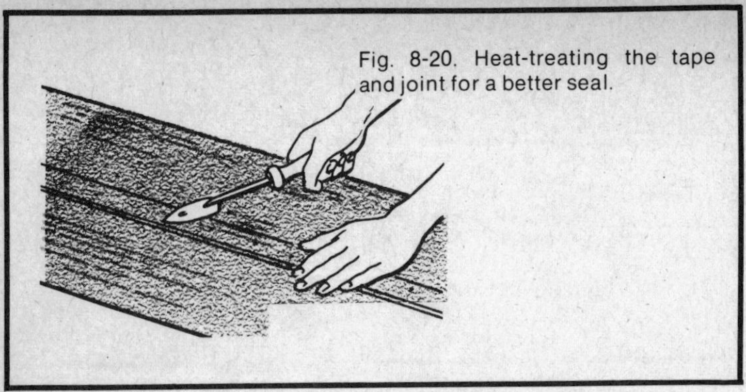

Fig. 8-20. Heat-treating the tape and joint for a better seal.

rub them down, then heat seal with a heat tool over the entire tape surface, making sure the heat seal application is uniform and thorough. These steps are illustrated in Figs. 8-17 through 8-20.

The remaining rectangular ductwork is installed just like these pieces; that is, fabricate regular pieces of duct as described in Figs. 8-5 through 8-15, providing transitions wherever the duct changes sizes.

The round branches are connected to the rectangular duct sections by using a round sheet-metal fitting the same diameter as the interior diameter of the round fiberglass duct. This fitting may be fabricated from standard sheet-metal pipe by snipping back one end with sheet-metal cutters to form the connecting tabs. At 90°, 180°, 270°, and 360°, cut deeper tabs for fitting against the outside of the duct as shown in Fig. 8-21. To cut the opening in the rectangular duct, use the fitting as a template, scribing the size of the cutout in the trunk duct. Fold

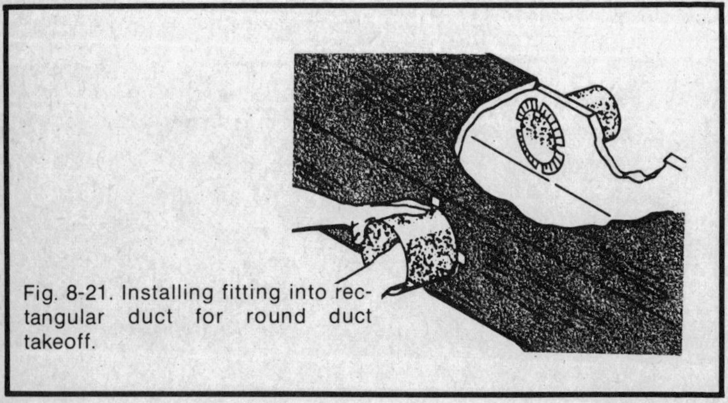

Fig. 8-21. Installing fitting into rectangular duct for round duct takeoff.

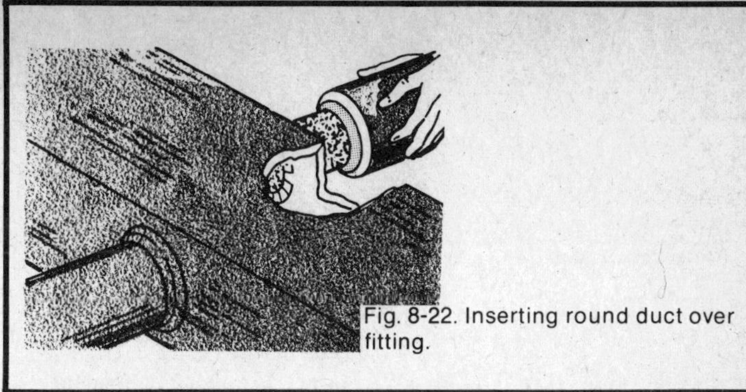
Fig. 8-22. Inserting round duct over fitting.

the four long tabs up; insert the fitting so the tabs rest flush agains the duct.

Complete the connection by sliding the round duct section over the metal sleeve so that it butts tightly against the rectangular duct section. Apply tape around the curvature of the connection,rub the tape down smoothly and apply heat as described previously (see Fig. 8-22).

Straight line connections are made in the round duct by using approved sheet-metal sleeve connectors to provide reinforcement and alignment, as shown in Fig. 8-23. Slide the round duct sections over the metal sleeve, butt tightly together, apply pressure sensitive tape, rub smooth and heat seal.

Elbows are constructed in round ducts by using standard sheet-metal elbows and two duct sections cut at a 45° angle. Use a template (usually provided with the round duct sections) and knife, or miter box and saw, to cut a 45° angle in the

Fig. 8-23. Method of joining two straight pieces of round fiberglass duct.

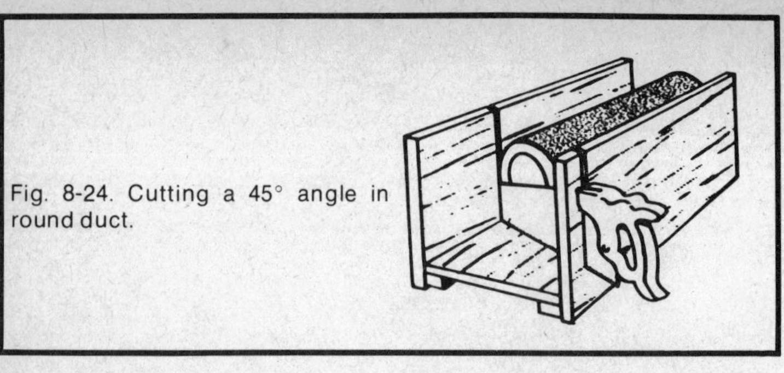
Fig. 8-24. Cutting a 45° angle in round duct.

section as shown in Fig. 8-24. Insert the sheet metal elbow (Fig. 8-25), making certain it is correctly aligned, and butt the ends of the sections tightly together over the elbow. Apply tape to the joint, rub down and heat seal as shown in Fig. 8-26.

Fig. 8-25. Joining two 45° sections of duct.

When joining two sections of round duct of different diameters, use a standard sheet-metal reducer as an internal reinforcement and alignment sleeve. Cut six "V" notches in the end of the larger duct section (Fig. 8-27) approximately 1-inch wide and equal in length to the tapered face of the metal

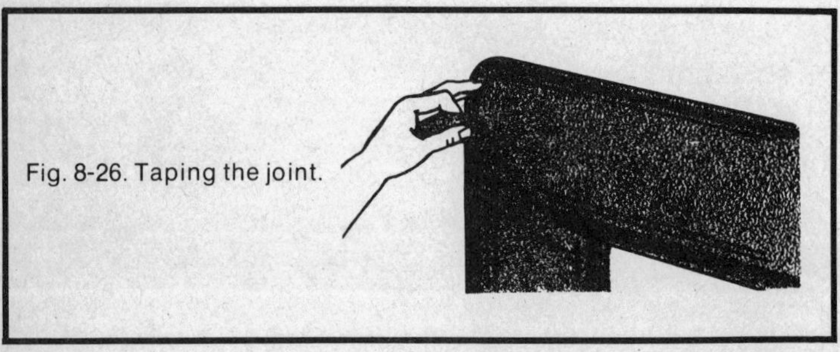
Fig. 8-26. Taping the joint.

Fig. 8-27. Method of reducing round ductwork by inserting sheet metal reducer into ducts.

reducer. Slip the reducer in the larger section of the duct until the tabs can be folded down against the tapered face and secure with pressure-sensitive tape around the circumference of the duct; also cover the "V" notches with tape as shown in

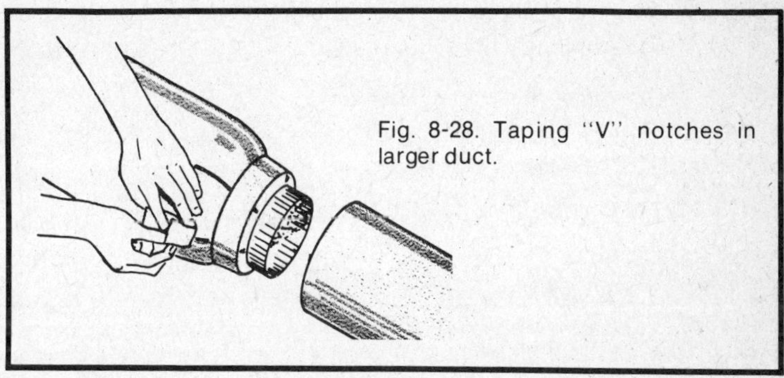
Fig. 8-28. Taping "V" notches in larger duct.

Fig. 8-28. Slip the smaller section of the duct over the reducer until it is butted tightly against the larger section, tape the joint securely, rub down the tape and heat seal (see Fig. 8-29).

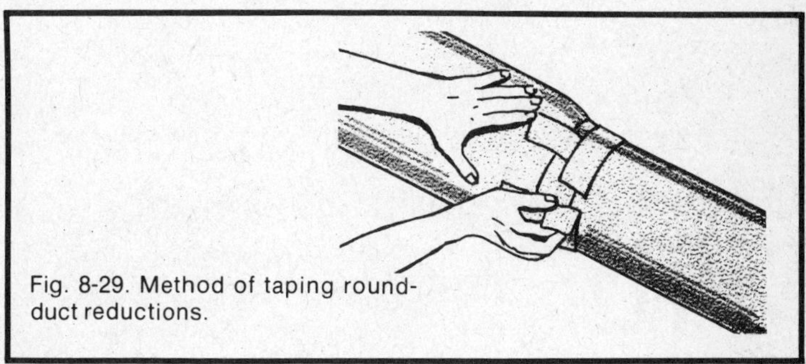
Fig. 8-29. Method of taping round-duct reductions.

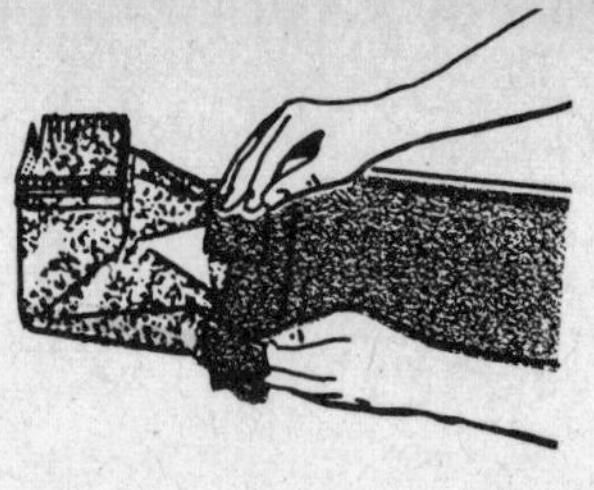

Fig. 8-30. Taping round duct to diffuser.

Supply run-outs may be installed on round ducts in a manner similar to those installed on rectangular ducts. Templates are also provided for this application.

Round duct is slipped onto round diffuser adapters and securely fastened to the sheet metal with two layers of tape. The tape should be applied so that one layer is on the sheet metal and one is on the duct, and overlapping each other by at least one inch as shown in Fig. 8-30.

A standard sheet metal pipe damper is easily adapted for use in round duct runs; dampers are highly recommended at each branch take-off for balancing the air supply to each area. Modification merely consists of installing an extended control arm so it will extend beyond the outer surface of the duct. Butt the two sections of duct together as shown in Fig. 8-31 and tape the joint securely with two layers of tape. Puncture the tape so the control arm may protrude before sealing the joint and installing the control handle.

The return air ductwork of the system in Fig. 8-3 requires the fabrication of an intake box over the return grille in the hallway of the residence. This box may be fabricated from flat

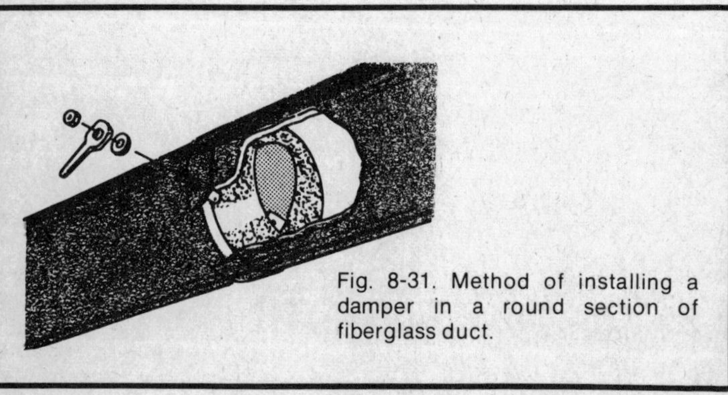

Fig. 8-31. Method of installing a damper in a round section of fiberglass duct.

duct board formed into rectangular sections with flush ends rather than shiplaps, capped at one end.

First measure the connecting collar to find the inside dimensions of the box or boot. Determine the stretchout dimensions in order to select the proper size of board, then measure and groove the board in the same manner as a one-piece duct section, omitting the shiplap at the end. Trim the section to the proper height and secure this section to the intake or outlet components with sheet-metal screws and washers. Drill pilot holes through the duct and register collar on 6-inch centers.

Next determine the size of the opening required for connecting the branch duct and make this cutout. If a round duct is to be connected, it will be necessary to install the metal connecting sleeve before capping the box. With the duct connection out of the way, the only remaining step is to fabricate and install an end cap on the box.

Measure the interior dimensions of the duct section to be capped, then select an adequate size duct board and mark these dimensions off on the insulation side, allowing an additional 2 1/2 inches on each side for the staple tab. Measure in one inch from the inside duct dimension marks and use a straightedge to make the cut.

Use the blue end cutoff tool to cut the stapled tabs. Prepare the tabs as described previously, removing the insulation from the board and scraping the jacket clean. Make a diagonal cut in each of the four corners to allow for folds in the staple tab.

Finally, install the end cap in the opening and attach by stapling the tabs on each side of the duct; be certain to pull the tabs tight before stapling to assure an air-tight closure. Apply tape and rub it thoroughly, then heat-seal.

SUSPENDING RECTANGULAR DUCT

Rectangular fiberglass duct requires suspension on eight-foot centers (maximum) for ducts up to 24-foot span; ducts spanning over 24 feet must be suspended at joints on four-foot centers. However, suspension may be eliminated if the building structural members provide adequate support within the above limitations. Any of the following suspension systems are satisfactory for this purpose.

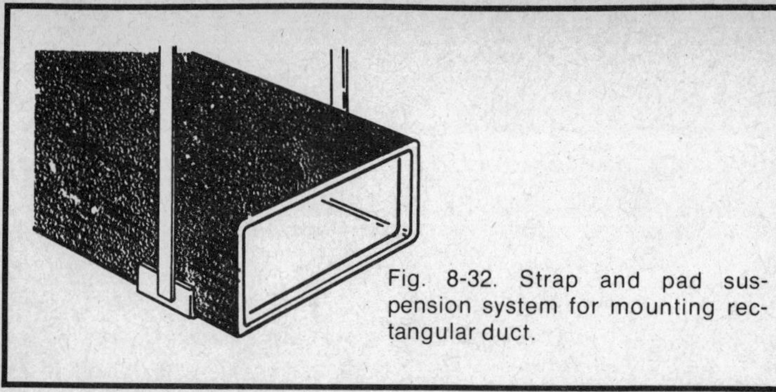

Fig. 8-32. Strap and pad suspension system for mounting rectangular duct.

1. *Strap and Pad System:* Suspend duct on 3/4-inch metal strap. Use 90° metal corner pads on duct corners to prevent jacket puncture (see Fig. 8-32).
2. *Strap and Channel Method:* Fig. 8-33 shows the ductwork suspended in a 3/4-inch metal strap which is in turn suspended from the framing and attached to a channel-iron member which passes under the duct.
3. *Strap and Rail System:* Suspend the duct on angle iron using 3/4-inch flat metal crossties spaced every three feet along the railing as shown in Fig. 8-34. Metal straps approximately 3/4-inch wide suspended from the roof, ceiling, or framing support the angles. Use sheet-metal screws for fastening the strap and rail members together.
4. *Strap and Angle System:* Suspend the duct on 3/4-inch metal strap hung from the building framing and attach angle iron passing under the duct as shown in Fig. 8-35.

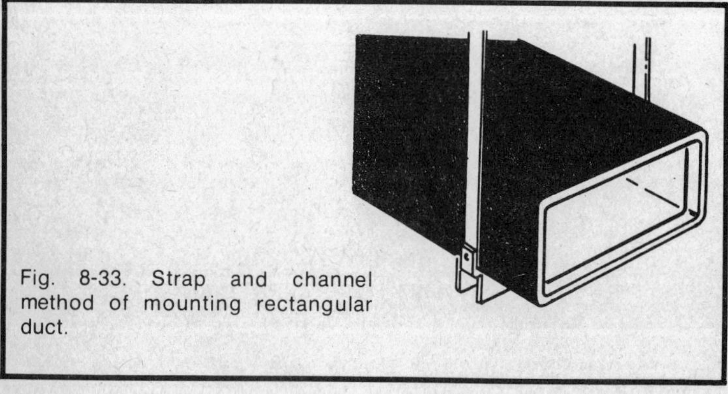

Fig. 8-33. Strap and channel method of mounting rectangular duct.

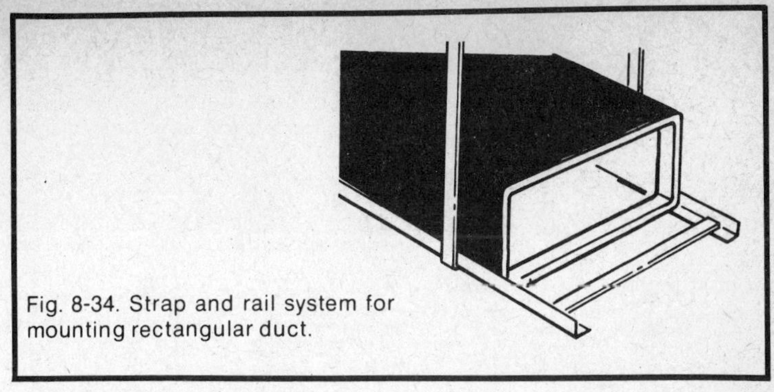
Fig. 8-34. Strap and rail system for mounting rectangular duct.

5. *T-Bar Saddle Method:* Suspend duct on standard stock T-bar (1 1/4-inch × 3/4-inch × 1 1/4 inches) as shown in Fig. 8-36. Trim back the 3/4-inch dimension or "T" and break it to 90° to form the T-bar saddle.

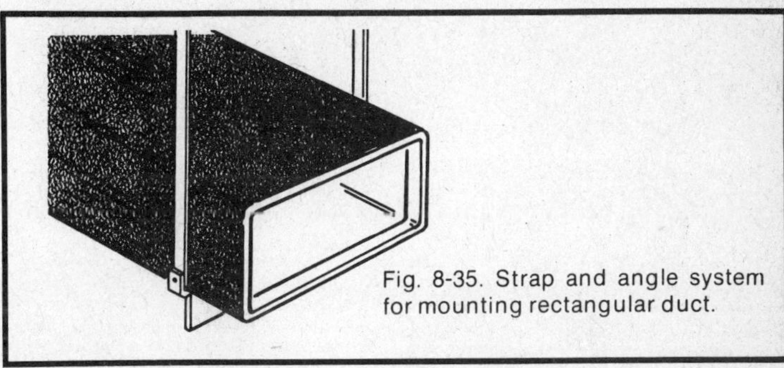
Fig. 8-35. Strap and angle system for mounting rectangular duct.

BRANCH TAKE-OFFS

In some cases it may be desirable to use air volume extractors constructed of sheet metal when making a branch take-off from the main duct or plenum. Figure 8-37 shows the

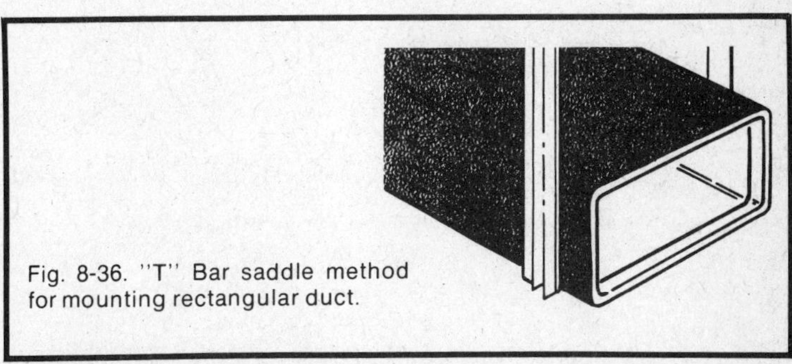
Fig. 8-36. "T" Bar saddle method for mounting rectangular duct.

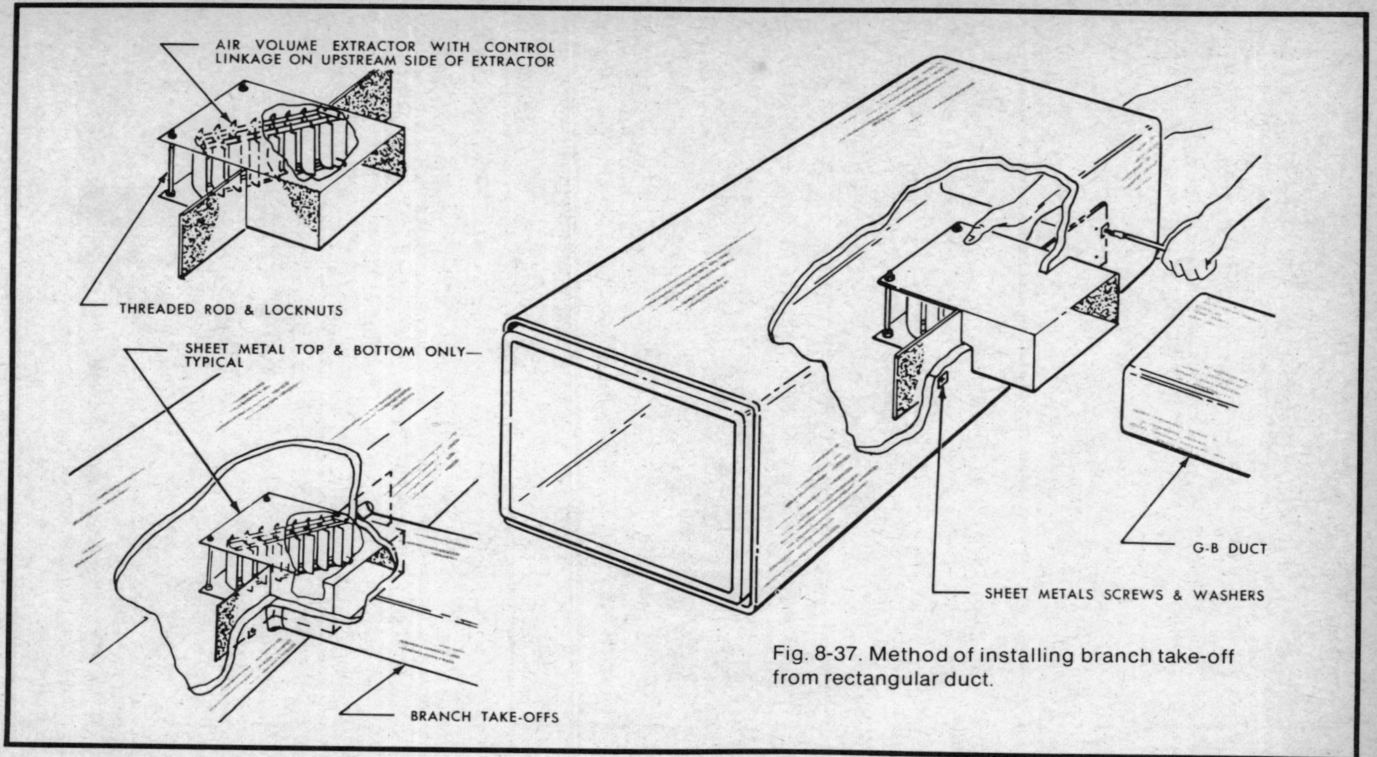

Fig. 8-37. Method of installing branch take-off from rectangular duct.

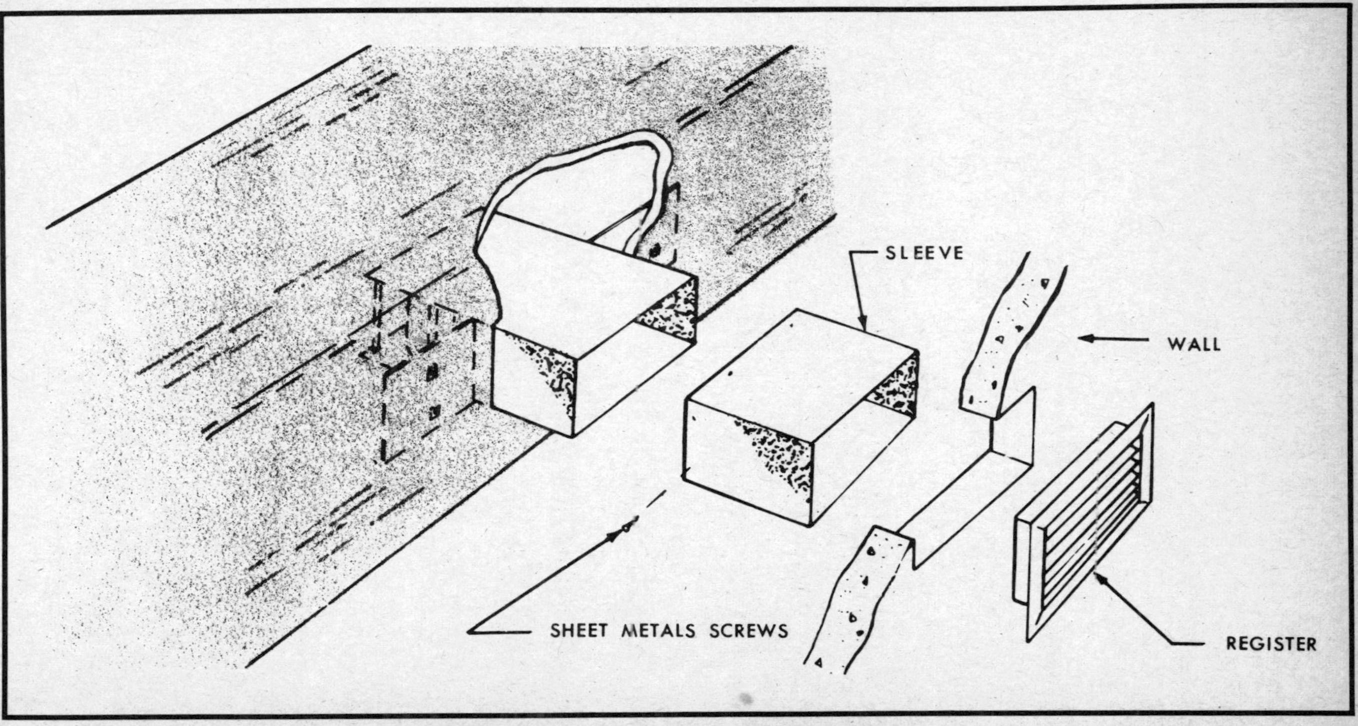

Fig. 8-38. Method of installing wall registers from rectangular duct.

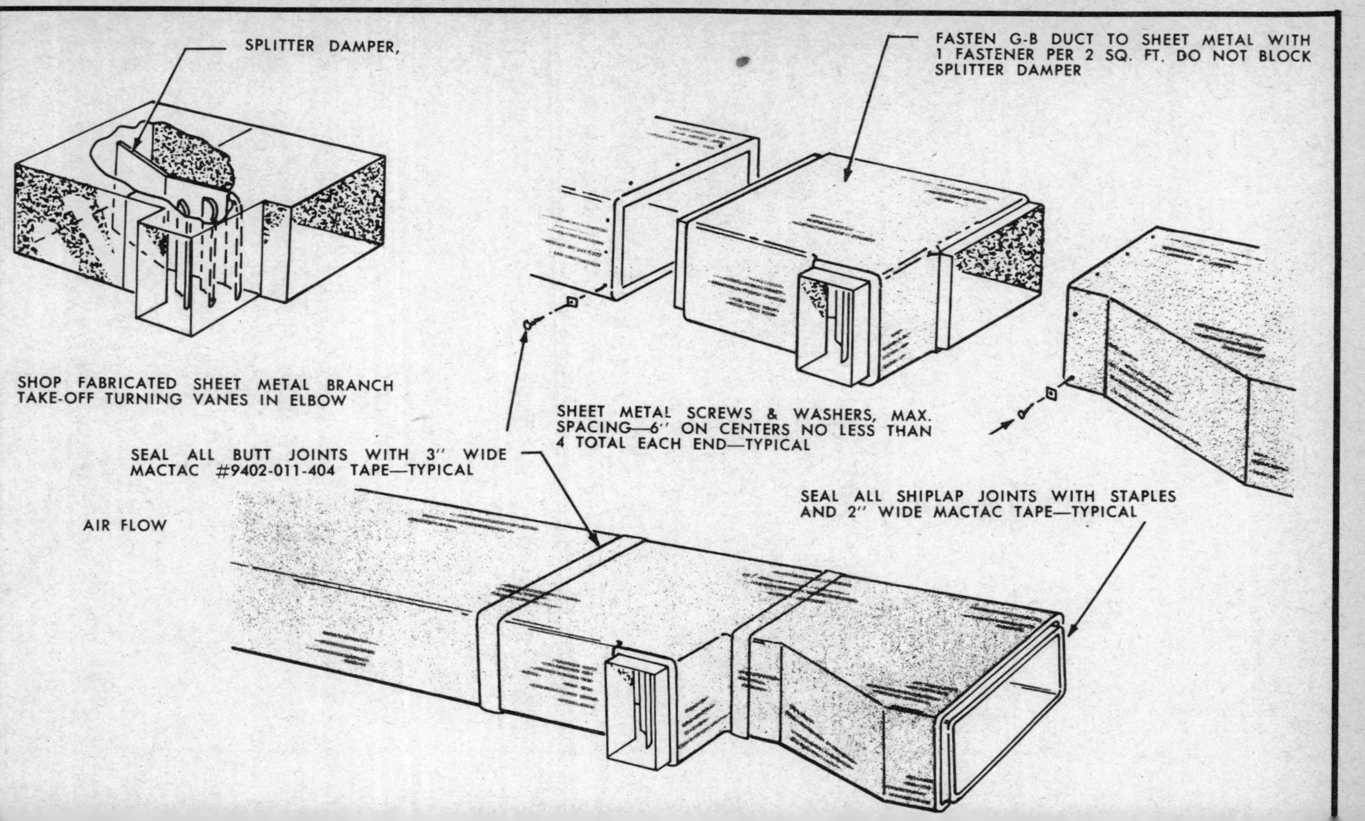

SPLITTER DAMPER,
FASTEN G-B DUCT TO SHEET METAL WITH 1 FASTENER PER 2 SQ. FT. DO NOT BLOCK SPLITTER DAMPER
SHOP FABRICATED SHEET METAL BRANCH TAKE-OFF TURNING VANES IN ELBOW
SHEET METAL SCREWS & WASHERS, MAX. SPACING—6'' ON CENTERS NO LESS THAN 4 TOTAL EACH END—TYPICAL
SEAL ALL BUTT JOINTS WITH 3'' WIDE MACTAC #9402-011-404 TAPE—TYPICAL
SEAL ALL SHIPLAP JOINTS WITH STAPLES AND 2'' WIDE MACTAC TAPE—TYPICAL
AIR FLOW

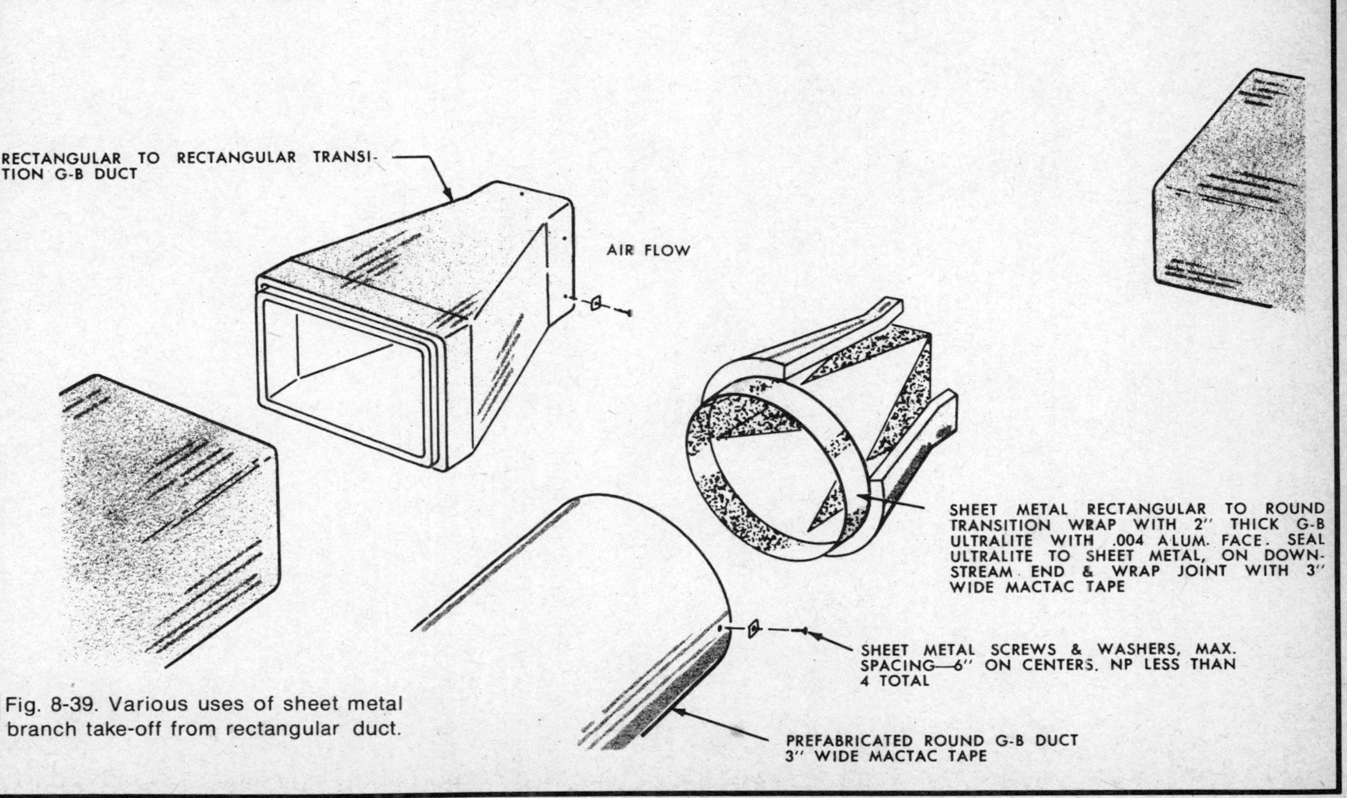

Fig. 8-39. Various uses of sheet metal branch take-off from rectangular duct.

installation of such an extractor in a duct take-off, while Fig. 8-38 shows the same extractor used in the installation of a wall register.

There are many other sheet-metal fittings available for various purposes in a duct system. A few of the branch take-off applications are shown in Fig. 8-39. Check with your local supplier of sheet-metal ductwork for help in planning and making up these various shapes.

Each manufacturer of fiberglass ducts may have a slightly different method of fabricating the system, but the procedures given in this chapter are typical. The main difference is nomenclature: one manufacturer may call his grooving tool a "Blue Tool," while another may call his the "No. 2 Tool," etc. The installation of fiberglass duct—especially the rectangular type—may seem too complicated for the average homeowner at first glance. However, once you make your initial mistakes and have that first section of duct made up, you'll find the installation is relatively easy, with the work going faster and faster as you progress.

The best part of all is that you will save not a little, but a lot of money in the process.

Chapter 9
Installing High-Velocity Systems

High-velocity heating/cooling systems are easily adapted to fit existing homes because they are so compact. Their use avoids much of the costly, time-consuming cutting and patching often necessary with conventional air-handling systems. So, if you're planning to add a central air conditioning system to your home which has no ductwork—or perhaps install a completely new comfort system—the high velocity system with its 3 1/2-inch diameter air ducts and 2-inch air outlets may be just the answer.

While there are several such systems on the market, Dunham-Bush, Inc., of Harrisonburg, Virginia, makes the Space-Pak kits especially for the do-it-yourselfer.

THE SYSTEM COMPONENTS

Figure 9-1 shows the basic components of a high-velocity residential heating and cooling system. Taking the individual items from left to right, we have:

1. *Return Air Assembly.* A centrally-located opening for this assembly's grille is probably the only major cutting required to install the system. An optional air cleaner is available which electrostatically removes dust particles from the air to keep your home clean.
2. *Blower Coil Unit.* Attic or basement installation of a high-velocity blower unit is compact and neat. Most

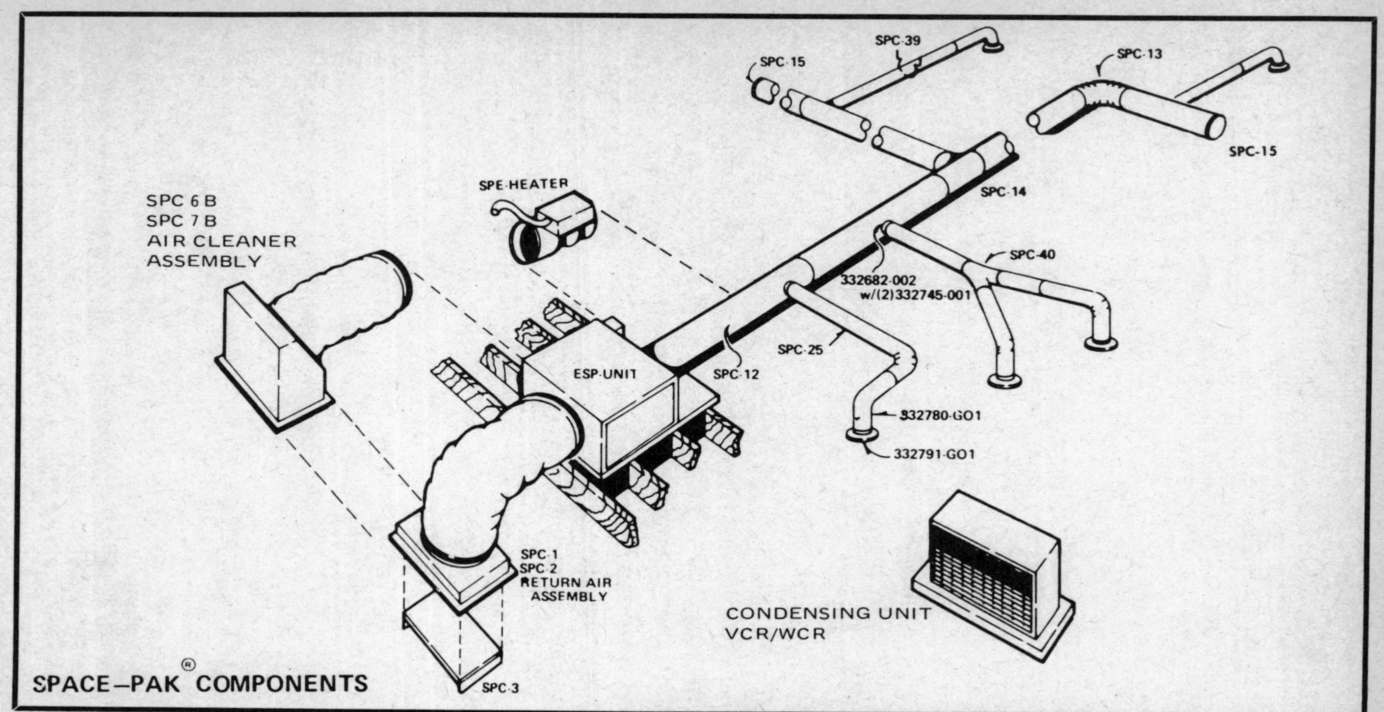

Fig. 9-1. Components of a high-velocity heating and cooling system.

kits include a spring-mounted, rubber-isolated blower and electric motor plus an expansion valve, condensate overflow control, and an extra-deep cooling coil that removes up to 30% more humidity from the air than shallow ones. Sizes are available from 17,000 to over 50,000 Btuh. A prefabricated seven-foot length of 14-inch or 18-inch flexible return air duct connects the return assembly to the blower (ESP) unit.

3. *Condensing Unit*. The condensing unit is the only piece of outside equipment needed. The twin blowers housed in an aluminum cabinet can be mounted as discussed in Chapter 7; that is, rooftop, through-the-wall, or on a concrete slab. The two-speed motor runs at low speed during normal operation for quietness, then automatically switches to high speed when the outside temperature is about 90°F. The quick-connect refrigerant lines (see Fig. 9-2) connect the condensing unit to the blower coil unit and can be installed in minutes.
4. *Duct Heater*. The optional electric heating system can be added in increments from 5 to 25 kilowatts or from 17,000 to 85,000 Btuh. The heating system is installed in the main plenum duct just downstream of the blower coil.
5. *Plenum Duct*. Seven-inch sound-absorbing fiberglass duct is used for the plenum. An aluminum vapor barrier on the outside of the plenum controls moisture; the flexible two-inch supply ducts are easily cut into the supply plenum.

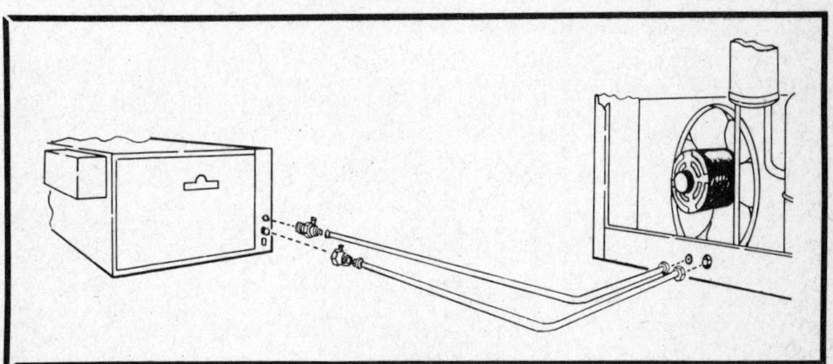

Fig. 9-2. Method of connecting refrigerant lines from condensing unit to fan coil unit.

6. *Air Supply Ducts.* Pre-insulated flexible supply ducts quickly connect to the main plenum duct; two parts join with a twist of the wrist, giving an airtight seal for good air delivery. The small (3 1/2-inch) ducts are easily run between studs and around obstacles, requiring less cutting and patching than conventional forced-air systems. An assortment of factory-made fittings—shown in Fig. 9-1—help speed the installation.
7. *Air Supply Outlet Plate.* The tiny two-inch opening will heat and cool the average room; it can be located in an out-of-the-way corner in ceiling or floor so that it is hardly noticeable. Adjustable dampers built into each outlet permit you to control the air flow according to the season or the number of people in the area.
8. *Controls.* Besides a conventional thermostat and control panel, the controls include: a factory-installed antifrost control which automatically shuts off the compressor if icing of the cooling coil occurs and a mercury float switch to prevent condensate overflow.

SYSTEM DESIGN

As with any system, proper sizing of a high-velocity system requires accurate calculation of heat gain and loss. Chapters 2 and 5 give the information necessary to perform these calculations.

System layout is a different story; only personnel trained in the design and application of high-velocity systems should attempt to lay one out. The calculations required are complicated and critical, and the numerous design considerations include the thermal load on, and the length of duct run to, each area.

So leave this job to the manufacturer's engineer, furnishing him with your home's floor plan and heat-loss and gain calculations. He will lay out the system completely, locating duct runs, return and supply fixtures, and all other system hardware at little or no cost to you. In addition, he will probably give you hints on installing the system and alert you to the most common mistakes.

Once the system is laid out to suit your residence, the small pre-insulated ducts and easily connected components make it possible to install the entire system in one day's time, although many installers like to spread the work over a period

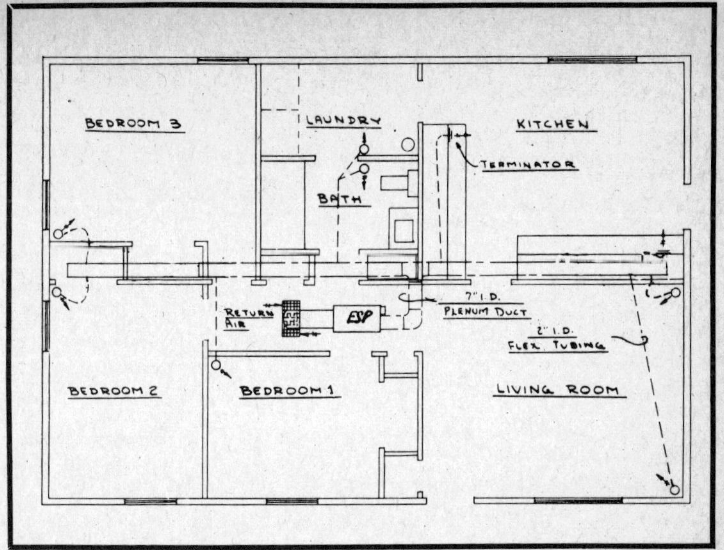

Fig. 9-3. Floor plan of a small residence with the high-velocity duct system laid out.

of two or three days. Fig. 9-3 shows a floor plan of a small residence with the high velocity system laid out.

BASIC INSTALLATION PROCEDURES

Many installers begin by locating the return air grille centrally in the home. Attic or basement location of the blower unit is typical; if there is no other access to your attic, the unit may be inserted through the opening cut for the return air grille.

When mounting the blower unit in the attic, it is recommended that vibration isolators be installed between the equipment and the house structure as shown in Fig. 9-4. With the unit in place, run the flexible return duct from the blower unit to the return air opening, making sure that this duct has a 90° turn in it with an inside radius of 6 inches or more. This is to prevent excessive noise in the return assembly. Insert the filter grille and snap the connecting bands in place for a neat installation.

The plenum duct can be placed in practically any location that will be accessible for attachment of the supply tubing. Your layout will show exactly how to run this duct, but the plenum is normally located in the attic or basement. In some

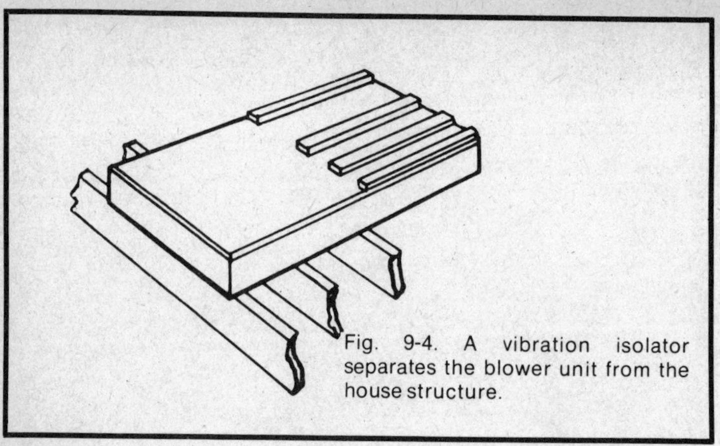

Fig. 9-4. A vibration isolator separates the blower unit from the house structure.

two-story or split-level homes it may be advantageous to run plenum to both levels, rather than make the smaller supply branches too long.

When boxing is necessary, as between floors or along the ceiling, the small size of the plenum makes it an easy job. (Figs. 9-5 and 9-6).

The 2-inch-inside-diameter (3 1/2-inch-outside-diameter) tubing conveys the conditioned air from the plenum to the air outlet or terminator. When installing this and any other ductwork, care should be taken to prevent sharp bends. The minimum radius for a tubing bend is 6 inches.

In two-story or split-level homes, supply tubing may be run from one story to another inside the walls without much difficulty. In older homes, however, hidden obstructions in the stud spaces can cause problems. In such cases it may be better to run the supply tubing from the attic down through second-story closets to the first floor (as shown in Fig. 9-7) or box the tubing in a corner (as shown in Fig. 9-8).

A critical factor in minimizing drafts with a high-velocity system is the location of the air outlets or terminators. In general, they should be out of the normal traffic patterns. This will prevent discharge air from blowing directly on the occupants.

Outlets should be located in the ceiling or floor with the center of the terminator 5 inches from the wall (or 5 inches from both walls in the case of a corner). They should never be located directly above shelves or large pieces of furniture. Neither should you attempt to install terminators for

Fig. 9-5. Boxing-in the plenum in a corner.

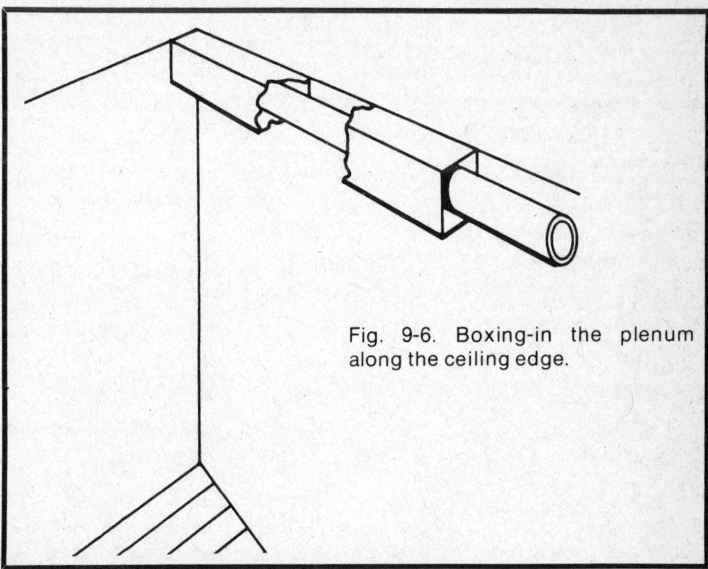

Fig. 9-6. Boxing-in the plenum along the ceiling edge.

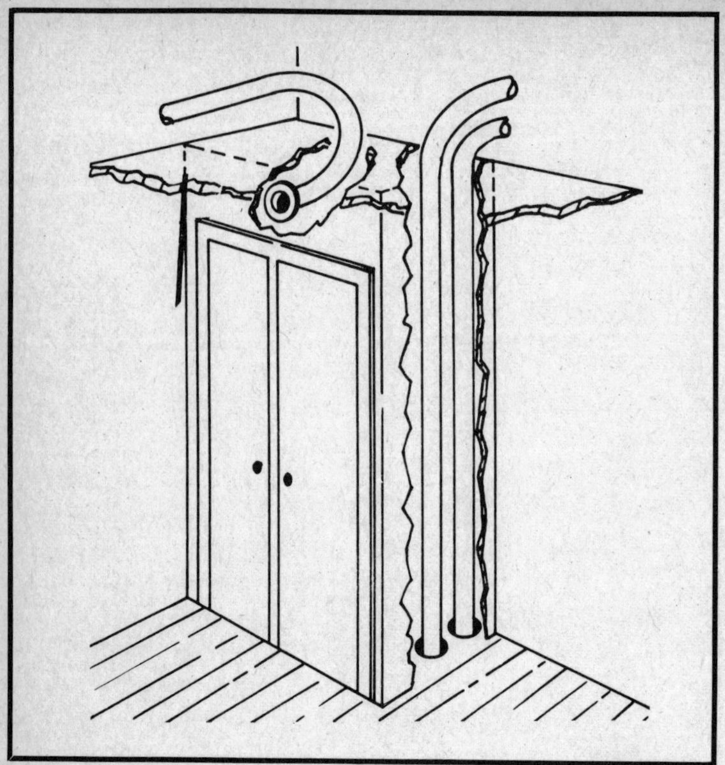

Fig. 9-7. Second-floor closets can be used to conceal tubing connecting the first floor and an attic unit.

horizontal discharge in wall partitions. The resulting sharp bend would probably be very noisy.

If you're installing the system for cooling only, the best location for the outlets is in the floor (Fig. 9-9) or ceiling (Fig. 9-10). Horizontal discharge for cooling is acceptable but is sometimes more difficult to install. Two excellent applications for horizontal discharge are the soffit above kitchen cabinets (Fig. 9-11) and the top portion of a closet (Fig. 9-7).

All comments regarding the location of outlets for cooling apply to heating also, with one exception. You should avoid horizontal air flow wherever possible because it does not maintain as narrow a floor-to-ceiling temperature differential as do floor and ceiling outlets.

An adjustable damper is supplied with each air outlet to balance the system. System balance for the heating function may differ from that for cooling. In fact a particular room may require more outlets for one season than for the other. If a given room requires three terminators for heating and only two for cooling, install three terminators; close the damper on one during the cooling season and open it during the heating season. However, you should remember that an open damper will still reduce the output of that duct by 10%, and when completely closed, it will reduce the output by only 80%.

With the blower coil, main plenum, branch ducts, and air outlets installed, the only remaining items are the condensing unit, the power wiring, and the controls.

The condensing unit is mounted outside the home on concrete pads or other supports as described in Chapter 7. The refrigerant tubing, and electrical wiring are also installed as described in that chapter.

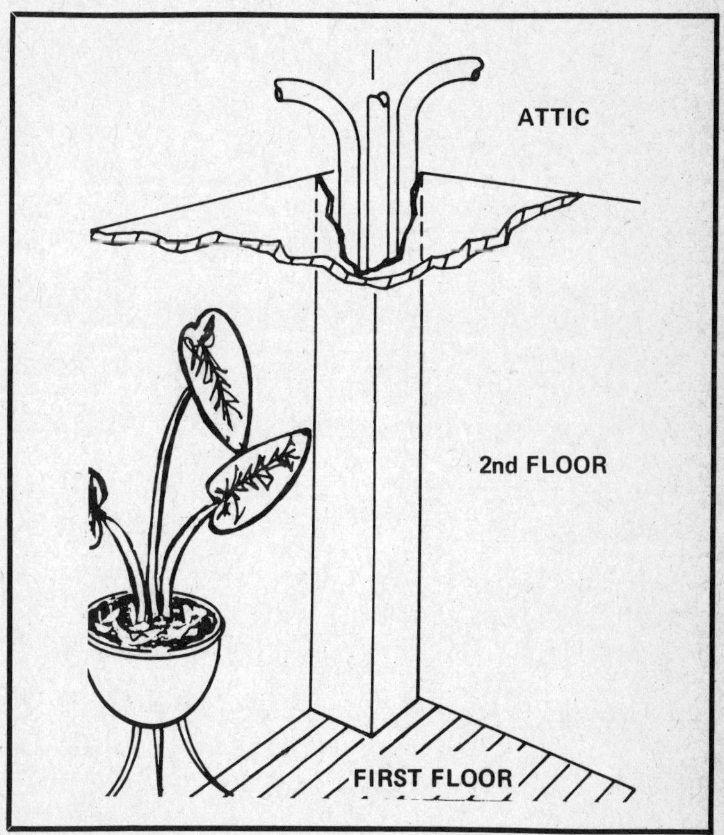

Fig. 9-8. Air-supply tubing from attic to first floor boxed in at a corner of the room.

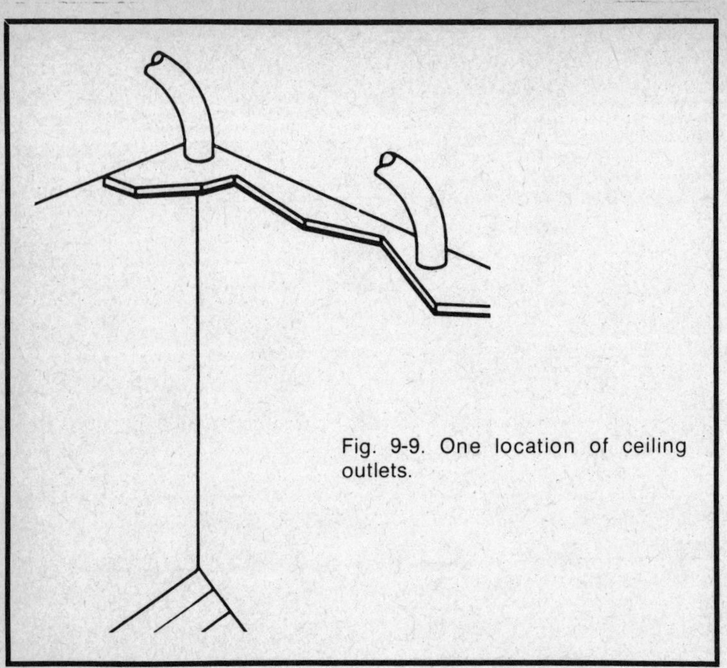
Fig. 9-9. One location of ceiling outlets.

Fig. 9-10. Proper location of a floor outlet in a corner.

Installations in a single-story residence (as shown in Fig. 9-3) are normally the quickest and simplest. Two-story applications require a little more planning and installation time, and are usually handled in one of two manners:

a. One blower unit and a plenum is installed in the attic of the home. Supply tubing is run from the plenum to (mostly) ceiling-mounted second-story outlets and down through the second-story walls to the first-story outlets, as shown in Fig. 9-12.
b. The second most common way to handle two-story homes is to install two separate systems (Fig. 9-13). One is placed in the attic to handle the second story using ceiling-mounted outlets and air return. The other system is placed in the basement or crawl space, with its duct system suspended from the floor joists and supplying floor outlets.

The two-system approach is well suited to homes with obstructions which would prevent in-wall duct routing. It usually enables a quick, trouble-free installation with an absolute minimum of cutting, patching, and duct runs, and

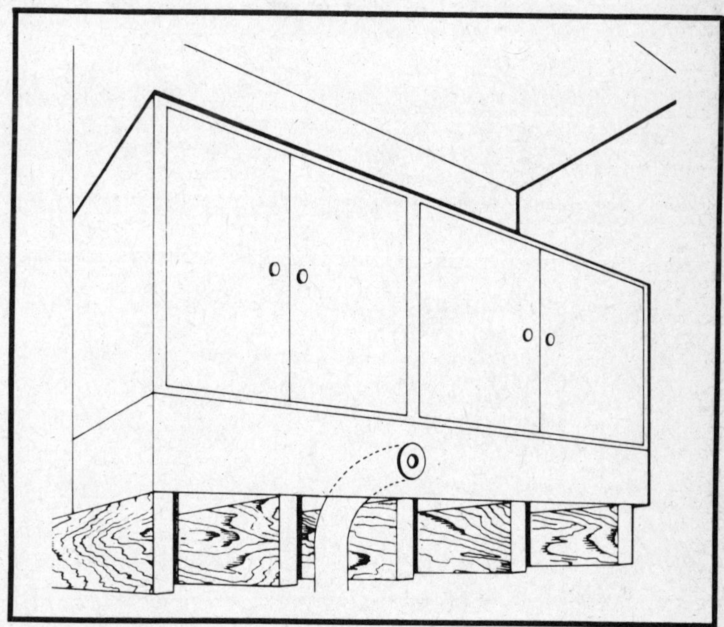

Fig. 9-11. Method used to terminate supply tubing in a kitchen soffit.

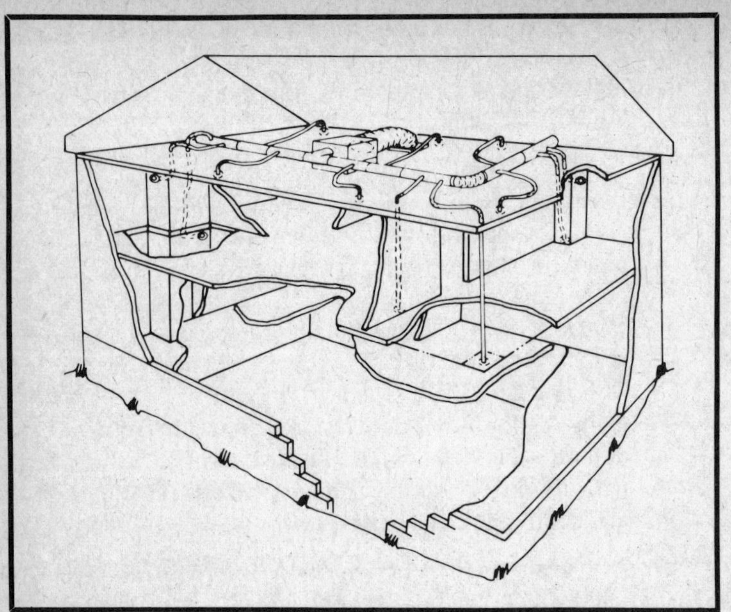

Fig. 9-12. One way to install a high-velocity duct system in a two-story house.

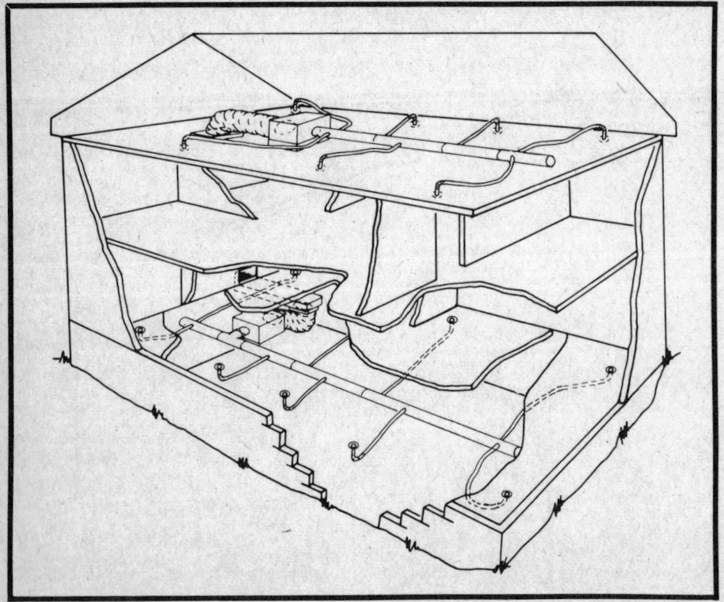

Fig. 9-13. Another method of conditioning a two-story house: two independent systems.

results in a zoned system with independent control of first- and second-floor climates.

Split-level homes are usually treated like a two-story home, with the blower installed in the attic, the return on the top level, and the supply tubing run down to the lower levels through walls, closets, or boxed in corners of rooms.

In some applications, however, it is more practical to go from one level to another with the plenum in order to hold down the length of supply duct runs.

Chapter 10
Tips on Conserving Energy

With the price of fuel climbing and reserves diminishing, energy conservation has become mandatory on both household and national levels. One way of saving on fuel bills is to survey your home and its heating and cooling system and see how you can cut down on wasted fuel. That crack around the back door lets expensive heated air out and cold air in during the winter; it could be repaired in 15 minutes for less than $1 if you'd take the time. How about your attic insulation? Chances are you could save as much as $300 a year—as you'll find out later—by investing $50 and about one hour of your time in insulation.

Those uninsulated heating ducts running through the attic space spend your good money to heat unused space. These are a few of the ways you can begin saving money immediately and help conserve energy at the same time.

Since Chapter 1 covered several ways of preparing your home for heating and cooling systems, including insulating, this chapter will deal with improving your heating and cooling system itself, as well as how to obtain the best service from it at the least cost.

PIPES AND DUCTS

Pipes and ducts are often ignored when considering energy-loss sources, and they shouldn't be. Insulation is as important on these heat transfer systems as it is in your walls and roof. Four excellent reasons for considering duct and pipe insulation are:

1. Reducing heat and cooling energy loss.
2. Maintaining proper water or air temperature.
3. Avoiding condensation problems.
4. Reducing heating effect on surrounding unused spaces.

Hot-water piping is usually insulated with either closed-cell plastic sponge insulation or asbestos fiber insulation. Ducts are normally insulated with glass fiber, polystyrene foam, or polyurethane foam.

Ductwork may not only be insulated externally or internally with fiberglass, but the ducts may also be made entirely out of fiberglass boards as described in Chapter 8 and shown in Fig. 10-1D. The remaining illustrations in Fig. 10-1 show examples of external and internal use of fiberglass board insulation and external blanket insulation. Application of the insulating material is extremely important: joints must fit tightly, and there must be no air gaps between the duct surface and the insulation. The insulation thickness will be dictated by the temperature difference between the medium in the duct

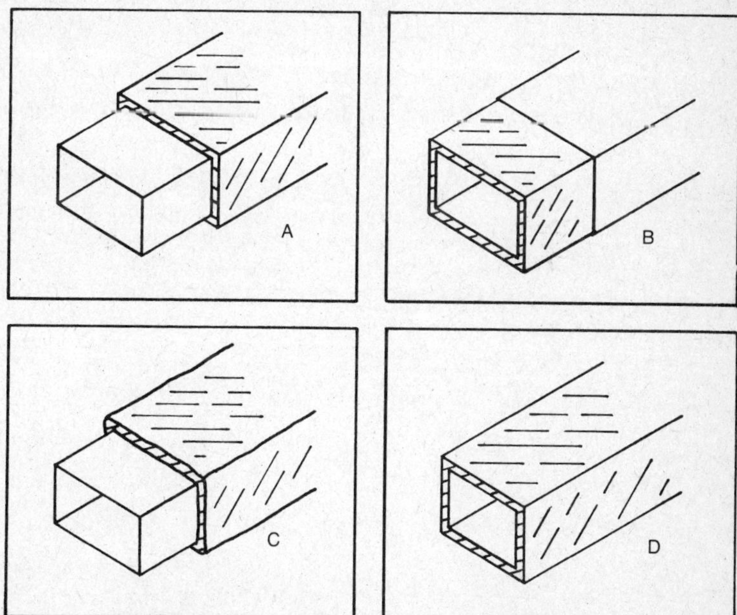

Fig. 10-1A. Sheet-metal duct with external fiberglass board insulation.
B. Sheet-metal duct with internal fiberglass board insulation.
C. Sheet-metal duct with an external blanket of fiberglass.
D. Duct fabricated from fiberglass board.

and the surrounding air; the greater the temperature difference, the thicker the insulation must be.

EQUIPMENT SIZING

For maximum comfort and efficiency, heating and cooling equipment for residential use should be sized to match the calculated heat gain and loss as closely as possible. People who guess at the size of the equipment they need or who aren't sure of their calculations sometimes use a "fudge factor" of 20–30%. This oversizing results in gross inefficiency as well as less-than-optimum comfort. Make absolutely sure that your equipment is the correct size for your home.

If, instead of a large central heat source, smaller multiple units are used, operation is usually more efficient. Nearly all electric furnaces, for example, have their heating elements wired in stages so that only one unit comes on at first. The remaining elements remain off until the first stage cannot maintain the temperature called for by the thermostat; then another stage automatically energizes, then another, and so on.

MOTORS

Matching a motor to the job it has to perform is another way to save energy. A motor that is too large or too small compared to the load it serves operates inefficiently. Replacing such units with the proper sizes will provide peak efficiency. Check the size of your fan motors (as well as those operating the circulating pumps if you have a hot-water system) to ascertain that they are of the proper size.

Low-voltage conditions also cause efficient motor operation and shortened life. Motor circuits should be sized properly to prevent excessive voltage drop (which should never exceed 3% of the applied voltage).

TURNING OFF UNNEEDED LIGHTS AND CLOSING OFF UNUSED ROOMS

Burning electric lamps needlessly not only wastes energy from the lamps themselves, but also causes your cooling system to work harder due to the extra heat put out by the lamps. You would probably be surprised at just how much you can save on your electric bill by turning off unneeded lights. You may possibly be able to save on needed light also. Lower the general illumination level of each area and locate small light fixtures by the work areas. This will lower both your electric and cooling bills.

While we're on the subject of light, during sunny winter days, open all the drapes to let in sunlight to help heat your home. At night, close the drapes for added insulation at your windows. In summer months the procedure is reversed.

Closing off unused rooms is another way to save on heating and cooling bills. How many rooms in your home are not used daily? One or two guest bedrooms? A workshop? If you insulate the walls of these areas to isolate them from the living area, then turn off the heat to them, you can save enough on your annual heating and cooling bill to take a week's vacation. If you are using individual units for heating and/or cooling in these areas, all you have to do is lower or shut off the thermostat controlling the areas. If the areas are conditioned by a central unit with ductwork, block off the air supply by means of dampers or similar devices.

It is not advisable to turn the heat off completely in unused areas in cold climates due to the possibility of frozen water pipes and structural damage. However, thermostats may be lowered to, say, 40 ° F, greatly reducing fuel bills without any harm to the home or its systems.

THERMOSTAT SET BACK

Opinions differ concerning the setting back of heating thermostats. Some maintain that it's best to leave your thermostat at 70°F twenty-four hours a day. Others insist that you can save energy by setting your thermostat back at night and turning it up the next morning. The first group maintains that the fuel consumed in the morning bringing the cold house up to a comfortable temperature offsets the savings that were realized during the night. The second group insists that savings are substantial despite the morning warm-up period.

Our experience has been that both groups are right! It depends entirely on the living habits of the occupants of the home, the type of heating system, and the condition of the home. A well-insulated home in which the occupants go to bed relatively early will use less fuel if its thermostat is set back at night. If the heating system is properly designed, it should bring the home's temperature up to 70°F quickly the following morning. You can even install a timer to shut off your heating

MAINTENANCE CHECK SHEET	WINDOW AIR CONDITIONER		**KEY TO SERVICE CHECKS**	O	Not applicable to this equipment
	CENTRAL AIR CONDITIONER			√	Component "OK" or Maintenance Service done
	HEAT PUMP			?	Component requires maintenance

CHECK LIST - HEAT PUMPS AND CENTRAL AIR CONDITIONING EQUIPMENT

AIR HANDLER EQUIPMENT		Lubricate motor bearings	
		Lubricate fan and pump bearings	
		Drive belts and alignment	
		Air filters—clean or replace	
		Indoor Coil—Air or water in ____ °F	
		Air or water out ____ °F	
		Conditioned space temp. ____ °F D.B.	
		Conditioned space temp. ____ °F W.B.	
		Clean Strainers	
		Leak test (ref. and water)	
		Drive motor volts ____	
		Drive motor amps. ____	
	ANNUAL		
		Clean fan blades and wheel vanes	
		Clean drip pan and drain **(condensate)**	
		Clean indoor coils **(liquid cleaner)**	
		Clean air intake grills or screens	
		Drain water at shut-down	
CONDENSER		Lubricate motor bearings	
		Lubricate fan and pump bearings	
		Drive belts and alignment	
		Spray nozzles and eliminators	
		Float valve	
		Bleed-off flow	
		Chemical water conditioner	

COMPRESSOR		Lubricate motor bearings	
		Motor drive belts or couplings and align.	
		Motor and comp. securing bolts	
		Crankcase oil level	
		Head pressure reading psi	
		Suction pressure reading psi	
		Oil pressure reading psi	
		Oil safety device	
		Hi-low pressure stat	
		Leak test (refrigerant and oil)	
		Crankcase oil heater - Amps ____	
	ANNUAL	Pump down and check suction and discharge read valves for leak back	
		Close shut-off valves	
		Comp. Volts & Amps.	
WATER COOLED CONDENSER		Water valve	
		Leak test (ref. and water)	
	ANNUAL	Drain at shut-down	
		Refill at start-up	
		Efficiency test (mineral deposits)	
		Thermostats	
		Magnetic contactors	
		Magnetic relays	
		Damper motors	

AIR COOLED AND EVAPORATIVE		
	Clean water strainer	
	Clean pump strainer	
	Clean air intake	
	Leak test (ref. and water)	
	Fan motor volts	
	Fan motor amps.	
	Pump motor volts	
	Pump motor amps.	
ANNUAL	Clean drip pan or sump	
ANNUAL	Drain at shut-down	
ANNUAL	Reconnect at start-up	
ANNUAL	Mineral or slime deposits on coil	
ANNUAL	Clean blower wheel vanes or fan blades	
ANNUAL	**Clean Condenser Coil (Liquid Cleaner)**	
ANNUAL		

CONTROLS	Solenoid valves	
CONTROLS	Thermostat expansion valves (super-heat setting, °F	
CONTROLS	B. P. and snap action valves	
CONTROLS	Defrost controls	
CONTROLS		
CONTROLS		
CONTROLS		
OTHER	Refrigerant quantity - sight glass	
OTHER	Duct damper position	
OTHER	Supplementary resistance heaters	
OTHER	Safety guards in position	
OTHER	Apply rust protective coating	
OTHER	Check for vibrating and rubbing piping	
OTHER	Clean external surface of all components with cloth	
OTHER	**Capacitor (Check for swelling)**	
OTHER		

CHECK LIST - WINDOW AIR CONDITIONER EQUIPMENT

CLEAN EXPOSED AIR SURFACES (Use steam cleaner or other recommended method)		
ANNUAL	Outdoor coil	
ANNUAL	Indoor coil	
ANNUAL	Fans and blower wheels	
ANNUAL	Motors and mountings	
ANNUAL	Cabinet and base	
ANNUAL	Grilles	
ANNUAL	Other components	
ANNUAL		
ANNUAL		
ANNUAL		

Apply rust protective coating	
Lubricate motor and fan bearings	
Install clean air filter	
TEST UNIT COOLING CAPACITY (after ½ hour's operating time)	
Room air temp.	_°F
Outdoor ambient temp.	_°F
Indoor coil air discharge temp.	_°F
Outdoor coil air discharge temp.	_°F
Test or check relay & thermostat function & contact condition	

MONTHLY	Install clean air filter	
MONTHLY		
MONTHLY		
MONTHLY		
MONTHLY		
OTHER		
OTHER		
OTHER		
OTHER		
OTHER		

Fig. 1C-2. Maintenance checklist.

system after you're in bed at night and turn it on in the morning before you awake. This way, you'll never notice the temperature difference, unless you have to get up during the night.

On the other hand, if family member have greatly varying sleeping habits—one goes to bed at 10 p.m., another at 11 p.m., another watches the late movie and stays up to 2 a.m., etc.—the family should probably leave their thermostat at 70°F and forget about it: very little could be saved by turning the thermostat back during the night.

We would advise each individual family to try a thermostat set-back for a month and then compare the fuel bill with last year's bill during the same month with no set-back. Providing the two months had similar weather, this will tell them if the practice is worthwhile in their home.

CLEANING EQUIPMENT

Keeping your heating and cooling equipment clean not only minimizes energy consumption and operating cost, but lengthens equipment life. Equally important is equipment maintenance: oiling motors, adjusting belts on fans, etc. Figure 10-2 will serve as a guide for cleaning and maintaining your heating and cooling equipment. Regular maintenance will keep your system in good running order, resulting in high efficiency and fuel savings.

Chapter 11
Troubleshooting Heating and Cooling Systems

The homeowner can save a great deal of time and money by doing his own troubleshooting and repairs when a problem occurs in his heating and cooling system. Even if he finds that the repair is too complicated for him, he will still probably save money by knowing what the problem is. This saves the repairman time, and gives the homeowner some basis for determining the fairness of the bill.

Troubleshooting heating and cooling systems covers a wide range of electrical and mechanical problems, from finding a short circuit in the power supply line, through adjusting a pulley on a motor shaft, to tracing loose connections in complex control circuits. However, in nearly all cases, the homeowner can determine the cause of the trouble by using a systematic approach, checking one part of the system at a time in the right order.

Every heating and cooling system's problems can be solved, and it is the purpose of this chapter to show the reader exactly how to go about solving the more conventional ones in a safe and logical manner.

The following data are arranged so that the problem is listed first. Then the possible causes of this problem are listed in the order in which they should be checked. Finally, solutions to the various problems are given, including step-by-step procedures where it is felt that they are necessary.

To better illustrate the use of these solutions to heating and cooling equipment problems, let's assume that your air-conditioner fan or blower motor is operating, but the compressor motor is not. Glance down the pages until you locate the problem titled "Compressor Motor and/or Condenser Motor will not start, but Blower Motor operates." Begin with item one under this title which tells you to check the thermostat system switch to make sure it is set to "COOL." Finding that the switch *is* set in the proper position, you continue on to item number two: "Check Temperature Setting." You may find that the temperature setting is above the room temperature so the system is not calling for cooling. Set the thermostat below room temperature, and the cooling unit will function.

This example is, of course, very simple, but most of your heating and cooling problems can be just as simple if you use a systematic approach to troubleshooting.

TYPICAL COOLING PROBLEMS

This section gives troubleshooting and repair procedures for common cooling-function problems.

Problem A:

COMPRESSOR MOTOR AND CONDENSER MOTOR WILL NOT START, BUT FAN/COIL UNIT (blower motor) OPERATES NORMALLY

(1) Check the thermostat system switch to ascertain that it is set to "COOL."
(2) Check the thermostat to make sure that it is set below room temperature.
(3) Check the thermostat to see if it is level. Most thermostats must be mounted level; any deviation will ruin their calibration. To correct, remove the cover plate, place a spirit level on top of the thermostat base (Fig. 11-1), loosen the mounting screws, and adjust the base until it is level; then tighten the mounting screws.
(4) Check all low voltage connections for tightness.
(5) Make a low-voltage check with a voltmeter on the condensate float switch: the condensate may not be draining. The float switch is normally found in the

fan/coil unit. A typical switch consists of a 24-volt, normally closed mercury switch attached to a stainless-steel arm and pin with a polystyrene float. The switch opens if the water level in the drain pan rises to approximately 1/2-inch. The water level in the drain pan should be between 1/4-inch and 3/8-inch when the unit is running.

In order to make the low voltage check, set the thermostat on cooling function and, with the fan running, check across the two lead wires at an accessible point (such as where they connect to a low-voltage terminal block). You should get a reading of 24 volts. If the float switch is broken, the mercury bulb is loose, the contact does not open or close, or the float is loose, replace the float switch assembly.

If you find the float switch to be in good working order, clean out the condensate drain and trap; then check to see if the fan/coil unit is level. Continue by checking the pitch (1/4-inch for each foot of horizontal line) of the condensate drain line.

(6) Low air flow could be causing the trouble, so check the air filters; if dirty, clean or replace.
(7) Make a low voltage check of the antifrost control; replace if defective.
(8) Check all duct connections to the fan-coil unit; repair if necessary.

Problem B:
COMPRESSOR, CONDENSER, AND FAN/COIL UNIT MOTORS WILL NOT START

(1) Check the thermostat system switch setting to ascertain that it is set to "COOL."
(2) Check thermostat setting to make sure it is below room temperature.
(3) Check thermostat to make sure it is level. If not, correct as shown in Fig. 11-1.
(4) Check all low-voltage connections for tightness.
(5) Check for a blown fuse or tripped circuit breaker. Determine the cause of the open circuit and then replace the fuses or reset the circuit breaker.
(6) Make a voltage check of low-voltage transformer; replace if defective.

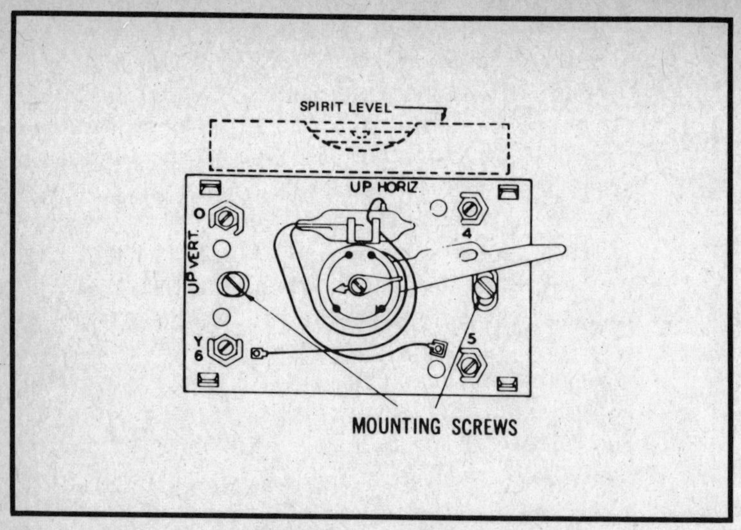

Fig. 11-1. Method of leveling a thermostat.

(7) Check your electrical service against minimum requirements; that is, for correct voltage, ampere rating, wire size, number of circuits, etc. Update your service if necessary.

Problem C:

CONDENSING UNIT CYCLES TOO FREQUENTLY, CONTACTOR OPENS AND CLOSES ON EACH CYCLE AND BLOWER MOTOR OPERATES

(1) The condensate drain may not be working properly. Run a check as described in No. 5 of Problem A. Then continue on to numbers 5, 6, 7, and 8 of the same problem.

(2) Check all low voltage wiring connections for tightness, and correct if necessary.

(3) The blower motor could be defective, so take an amperage reading on the motor while it is running. However, do not confuse the full load (starting) amperes shown on the motor rating plate with the actual running amperes. The latter should be about 25% less. If the amperage varies considerably from that on the nameplate, have the motor checked for bad bearings, defective winding insulation, etc. The problem could also be caused by low supply voltage.

Problem D:

INADEQUATE COOLING WITH CONDENSING UNIT AND BLOWER RUNNING CONTINUOUSLY

(1) Check all low-voltage connections (control wiring) against the wiring diagram furnished with the system. Correct if necessary. Then check for leaks in the refrigerant lines.

(2) Check all joints in the supply and return ductwork; make all joints tight.

(3) The equipment could be undersized. Check heat gain calculations against the output of your unit. Correct structural deficiencies with insulation, awnings, etc., or install properly sized equipment.

Problem E:

CONDENSING UNIT CYCLES BUT BLOWER MOTOR DOES NOT RUN

(1) Check all low voltage connections against the wiring diagram furnished with the system. Correct if necessary.

(2) Check all low voltage connections for tightness.

(3) Make a voltage check on the blower relay and replace if necessary.

(4) Make electrical and mechanical checks on blower motor. Check for correct voltage at motor terminals. Mechanical problems could be bad bearings or a loose blower wheel. Bearing trouble can be detected by turning the blower wheel by hand (with current off), and checking for excessive wear, roughness, or seizure.

Problem F:

CONTINUOUS SHORT CYCLING OF BLOWER COIL UNIT AND INSUFFICIENT COOLING

(1) Make electrical and mechanical checks as described in No. 4 of Problem E. Repair or replace motor if necessary.

Problem G:

SWEATING AT BLOWER COIL OUTPUT OR AT ELECTRIC DUCT HEATER OUTLET

(1) Check to see if the insulation is installed properly.

(2) Inspect the joints at the duct heater or blower coil receiving collar; seal properly.

TYPICAL HEATING PROBLEMS

This section details diagnosis and repair procedures for problems specific to the heating function.

Problem H:
THERMOSTAT CALLS FOR HEAT BUT BLOWER MOTOR WILL NOT OPERATE

(1) Check all low-voltage connections against the wiring diagram furnished with the system. Correct if necessary.
(2) Check all low-voltage connections for tightness.
(3) Check line voltage against the unit's nameplate. Correct if necessary.
(4) Check all line-voltage connections for tightness.
(5) Check for blown fuses or a tripped circuit breaker in the line; determine the reason for the open circuit and replace the fuses or reset the circuit breaker.
(6) Check the low-voltage transformer; replace if defective.
(7) Make a low-voltage check on the magnetic relay; repair or replace if necessary.
(8) Make electrical and mechanical checks on the blower motor as described in No. 4 of Problem E (under cooling). Repair or replace the motor if defective.

Problem I:
THERMOSTAT CALLS FOR HEAT, BLOWER MOTOR OPERATES BUT DELIVERS COLD AIR

(1) Make a visual and electrical check on the heating elements. Replace if defective.
(2) Make an electrical check on the heater limit switch—first disconnecting all power to the unit—using an ohmmeter to check continuity between the two terminals of the switch. If the limit switch is open, replace it with one of the same rating.
(3) Make an electrical check on the time-delay relay. Most are rated at 24 volts and have one set of normally-open main contacts for line duty and one set of normally-open auxiliary contacts for pilot duty. First check the voltage between terminals H^1 and H^2 with the relay heater energized. The reading should be 24 volts. Next connect the voltmeter across terminals

M^2 on the time-delay relay and L^2 on the terminal block. After a delay of about 45 seconds there should be full line voltage across these contacts and the heating element should be energized. Again, use the voltmeter and check the voltage across terminals A^2 and H^2 on the time-delay relay. This should read 24 volts and will indicate if the pilot duty contacts have closed with the line duty contacts. If they do not close after the relay heater has been energized, disconnect the power to the unit and check for continuity between terminals H^1 and H^2 with an ohmmeter. Then check for continuity between each terminal and ground. If the relay heater coil is open or grounded, replace the time-delay relay. If the time-delay relay heater checks OK and the main and auxiliary contacts do not close, this indicates the cross bar is binding or misaligned, and the time delay should be replaced.

(4) Make an electrical check on the magnetic relay and repair or replace if defective.

(5) Check your electrical service entrance and related circuits against the minimum recommendations.

Problem J:

THERMOSTAT CALLS FOR HEAT AND BLOWER MOTOR OPERATES CONTINUOUSLY. SYSTEM DELIVERS WARM AIR BUT THE THERMOSTAT IS NOT SATISFIED

(1) Check all joints in the ductwork for air leaks, making all defective joints tight.

(2) Check all duct joints and blower outlets for tightness and seal where necessary.

(3) Make a visual and electrical check of the electric heating element and repair or replace if necessary.

(4) Make an electrical check on the heater limit switch as described in No. 2 of Problem I.

(5) Make an electrical check on the heater limit switch as described in No. 3 of Problem I.

(6) Check the heating element against the blower unit for the possibility of a mismatch.

(7) Check your heat loss calculations. The equipment could be undersized. If so, correct structural deficiencies by installing more insulation, storm windows and doors, etc., or install properly sized equipment.

Problem K:

BLOWER UNIT OPERATES PROPERLY AND DELIVERS AIR BUT THERMOSTAT IS NOT SATISFIED

(1) Check all joints in the ductwork for air leaks and repair if necessary.

(2) Check the air filter and clean or replace if necessary. Also check the number of air outlets for adequacy and make sure they are balanced.

(3) Check for undersized equipment as described in No. 7 of Problem J.

Problem L:

ELECTRIC HEATER CYCLES ON LIMIT SWITCHES BUT BLOWER MOTOR DOES NOT OPERATE

(1) Make an electrical check on the magnetic relay and repair or replace if defective.

(2) Make electrical and mechanical checks on the blower motor as described in No. 4 of Problem E (under cooling).

(3) Check the line connection against the wiring diagram furnished with the system.

TYPICAL HEATING & COOLING PROBLEMS

This section gives troubleshooting and repair procedures for problems affecting both system functions.

Problem M:

EXCESSIVE AIR NOISE AT TERMINATOR

(1) Duct or outlet undersized; air velocity too great. Increase size of duct and/or outlet.

(2) Make external static pressure check and correct restrictions in system if necessary.

(3) Check for properly balanced system and make corrections if necessary.

Problem N:

EXCESSIVE NOISE AT RETURN AIR GRILLE

(1) Check the return duct to make sure it has a 90° bend in it as shown in Fig. 11-2.

(2) Make a visual check of the blower unit to ascertain that all shipping blocks and angles have been removed.

(3) Check blower motor assembly suspension and fasteners and tighten if necessary.

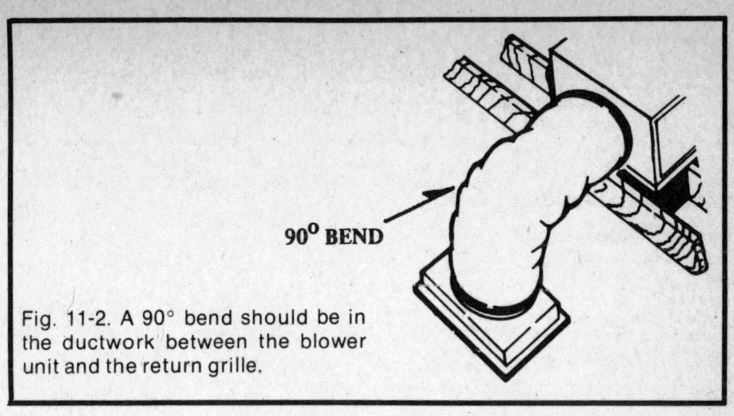

Fig. 11-2. A 90° bend should be in the ductwork between the blower unit and the return grille.

Problem O:

EXCESSIVE VIBRATION AT BLOWER UNIT

(1) Visually check for vibration isolators (which isolate the blower coil from the structure). If missing, install as recommended by the manufacturer.
(2) Visually check to ascertain that shipping blocks and angles have been removed from the blower unit.
(3) Check blower motor assembly suspension and fasteners and tighten if necessary.

MAINTENANCE OF HEATING & COOLING EQUIPMENT

The old saying, "an ounce of prevention..." certainly holds true for heating and cooling systems. A correctly installed system that is maintained according to the manufacturer's recommendations will give you years of trouble-free service at minimum cost.

Maintenance procedures and frequency will depend on the type of system you have, but the information given here covers most residential heating and cooling systems in general. For further information, consult the Owner's Handbook accompanying your equipment. If you don't have one, ask your supplier or write directly to the manufacturer to obtain a copy.

Besides saving on repairs, good maintenance of your heating and cooling equipment will ensure that your equipment operates at maximum efficiency, which will save fuel and reduce other operating expenses.

Cleaning

Heating and cooling equipment should be cleaned at regular intervals in order to maintain operating efficiency, lengthen the life of the equipment, and minimize energy consumption and operating costs.

Begin by removing lint and dust with a cloth and brush from all finned convector-type heaters, and then vacuum them. This includes electric and hot-water baseboard heaters, wall, and unit heaters. Air conditioning evaporator and condenser coils should be vacuumed, scrubbed with a liquid solvent or detergent, and flushed.

Air filters in forced-air systems collect dust, as is their purpose. Periodic inspection will tell you how often they should be cleaned or replaced. Permanent metal mesh and electrostatic air filters should be washed and treated; throw-away filters should be replaced with the same size and type. Obtain them from your local hardware store or mechanical contractor.

Vacuum or wipe off lint and dust from all supply and return grilles, diffusers, and registers, using a detergent solution if necessary. While you're doing this, also remove any dirt on the dampers and check the levers for proper operation.

Motors should also be vacuumed to remove dust and lint, but you should first turn off the power supply. Once free of dust and lint, wipe their exterior surfaces clean with a rag and reconnect the power.

Propeller fans and blower wheels are especially susceptible to dust deposits and should be cleaned often. Again, make certain that the power is turned off—to prevent losing a finger—and remove the dirt deposits with a liquid solvent.

Every periodic inspection of your heating and cooling system should include a check of the condensate drain. Wash the drain pan with a mild detergent and flush out the drain line. All loose particles of dirt should be brushed from the evaporator coil and a fin comb should be used to open all clogged air passages in the coil. If the coil is extremely dirty, a small pressurized sprayer may be used with a strong dishwasher detergent to flush the coil. Always rinse with clean water after using detergent.

Lubrication

Adequate, regular lubrication ensures efficient operation, long equipment life, and minimum maintenance cost, but never over-lubricate! Observe the manufacturer's instructions when lubricating all bearings, rotary seals, and movable linkages of:

1. Motors—Direct or belt-drive type.
2. Shafts—Fans, blower wheel, and damper.
3. Pumps—Water Circulating.
4. Motor Controllers—Sequential and damper operators.

Periodic Inspection

This check list should be followed during periodic inspections of your heating and cooling system in order to minimize maintenance expense and to save as much fuel as possible.

1. *DRIVE BELTS*

   Examine for:

   Proper tension and alignment.
   Sidewall wear.
   Deterioration, cracks.
   Greasy surfaces.
   Safety guards in position and secure.

2. *"V" PULLEYS*

   Examine for:

   Alignment.
   Wear of "V" wall.
   Tightness of pulley and set screws.

3. *FAN BLADE AND BLOWER WHEEL ASSEMBLIES*

   Examine for:

   Metal fatigue cracks.
   Tightness of hubs and set screws.
   Balance.
   Safety guards in position and secure.

4. *ELECTRICAL COMPONENTS*

   Examine, burnish, or replace electrical contacts of:

   Magnetic contactors and relays.
   Thermal relays.
   Control switches, thermostats, timers, etc.

   Examine wire and terminals for:

   Corrosion and looseness at switches, relays, thermostats, controllers, fuse clips, capacitors, etc.

   Examine motor capacitors for:

   Case swelling.
   Electrolyte leakage

When checking the electrical system examine all components for evidence of overheating and insulation deterioration.

Chapter 12

Install An Attic Fan: Added Comfort and Lower Air-Conditioning Bills

A ventilating fan installed in the attic of your home can lower attic temperatures as much as 40°F, resulting in a much cooler house during the hot summer months. This means that your air-conditioning system will not have to operate as much, which in turn lowers your utility bills considerably.

A high-capacity attic fan can also cool the entire house. Such a system pulls in cool outside air, exhausts hot inside air, and lowers room temperatures 10° to 20°F. In additon to lowered temperatures, such a system provides a constant flow of fresh air for easier breathing and greater comfort.

For example, during a hot day, the sun beating down on your unventilated attic can bring the temperature up to 130° or more. In summer the attic will seldom drop to lower than 110°F, even during the night. At night the outside temperature may drop to 75°F or even lower, but the hot attic keeps the temperature of your home's interior to as high as 95°F. If you install a ventilating fan, the cool outside air is drawn in through the windows, circulated through the house, and exhausted through louvers in the attic. This lowers both the attic and room temperatures considerably.

The equipment and operating cost of a home ventilating system is only a fraction of the cost of central air conditioning. Some people even prefer the "feel" of a fan-cooled house to one that is completely air-conditioned. In some areas, air conditioning is almost mandatory during summer days, but at night, in almost any location, an attic fan will cool the home sufficiently.

SIZING THE FAN

The size of fan necessary to properly ventilate a home depends on the size of the house and also the number of air changes desired each minute. Check Fig. 12-1 to see how many air changes per minute are recommended for your location. If you live in Zone 1, you'll need one complete air change per minute; in Zone 2, one change per 1 1/2 minutes will suffice.

To illustrate the process of sizing a fan, let's assume that a one-story house measures 30 × 40 feet and has an eight-foot ceiling. The volume of the house is 30 × 40 × 8 or 9600 cubic feet. If one air change per minute is required, the ventilating fan must move 9600 CFM (cubic feet per mintue) of air. For 1 air change every 1 1/2 minutes, multiply the volume of the house by 0.66. In the previous example, the ventilating fan must handle 9600 (cubic feet) × 0.66 or 6400 CFM. Or you can use the following chart to determine the size fan you need for your home.

First calculate the volume of your home, then find the closest figure in the proper zone column in the chart below to determine the size of attic fan you'll need. When a flow of more than 16,000 CFM is required, use two or more fans.

Fan Size	CFM	Zone 1 Volume of house	Zone 2 Volume of house
24-inch	5200	4000–5250	6000–7900
30-inch	7200	5400–7000	8700–10800
36-inch	10550	10000–12000	15000–18000

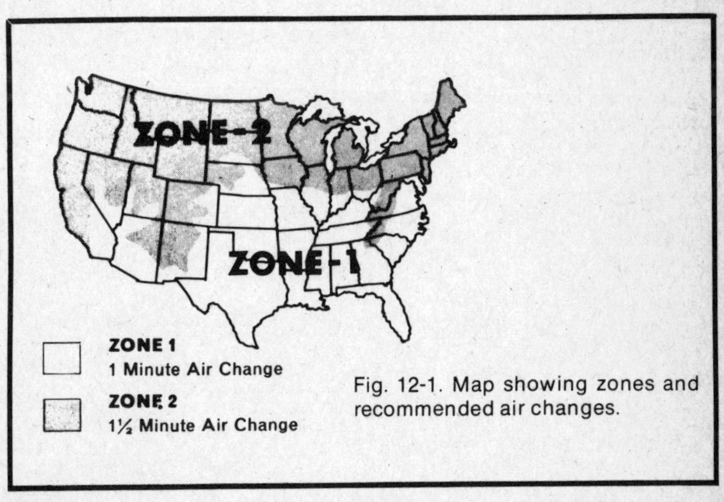

Fig. 12-1. Map showing zones and recommended air changes.

Fig. 12-2. Elevation view of a one-story house with an attic ventilating fan located in the hall.

The fan should be located centrally in the house and above the ceiling to provide the ideal flow of air. Figure 12-2 shows an elevation view of a one-story house with ventilating fan located in the hall. The fan is mounted above automatic shutters. Exhaust vents are located in the gable ends of the roof.

Such a fan should be placed below the attic rafters a distance at least equal to its own diameter. Therefore, a fan 30 inches in diameter should be placed at least 30 inches below the attic rafters. This distance will permit the air to flow freely through the fan, whereas closer spacing may cause the fan to overwork and be unable to discharge the amount of air for which it is rated. The fan motor will also become hot because of this extra work and probably burn out in a very short period of time.

VENTS

Adequate exhaust vents for the fan discharge must also be provided; the size of these vents is governed by the following factors:

1. Obstruction in the vents: mesh screening, hardware cloth, or none.
2. Whether the louvers are constructed of wood or metal.
3. The thickness of the louvers.

The gable ends of the attic, as shown in Fig. 12-3, are good locations for exhaust grilles. If only one gable end is used, the exhaust grille should be placed at the end away from the prevailing winds. On hip roofs, louvers may be built into dormers of various types, such as the one shown in Fig. 12 1. If

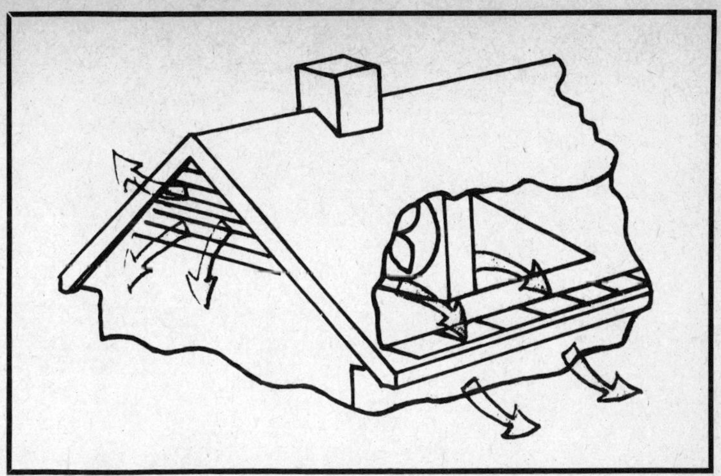
Fig. 12-3. Exhaust grilles at the gable end of a roof and under the eaves.

the gable opening of a house is not sufficient for exhausted air, screened openings in the overhanging eaves may be provided.

Houses or cabins with flat or shed roofs present a problem in housing the fan as well as discharging the air. A wall or roof exhaust fan may be the best solution in this case (see Figs. 12-5 and 12-6). However, a penthouse or cupola can be built onto the roof to house the exhaust fan and louvers.

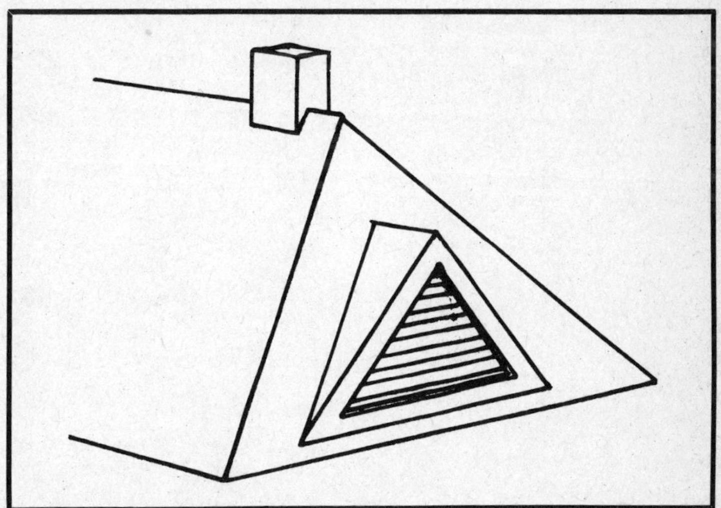
Fig. 12-4. An exhaust dormer built into a hip roof to house an exhaust grille.

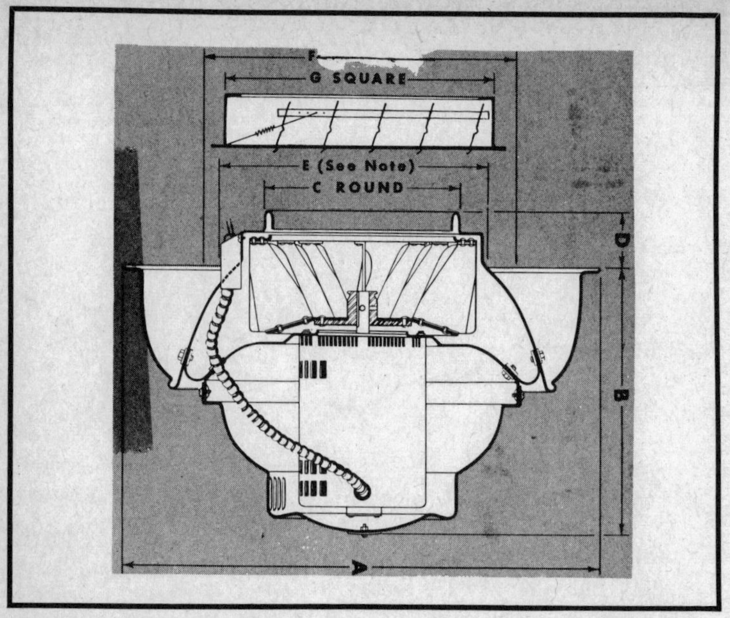

Fig. 12-5. A wall exhaust fan suitable for ventilating buildings with flat or shed roofs.

The louver openings must be large enough to accomodate the fan size selected for proper operation. The following table lists the approximate size of exhaust opening needed for various types of openings and sizes of exhaust fans:

Type And Size of Opening (square feet)

Fan Size (inches)	Full Opening	Wood Louvers	Metal Louvers
22	11	16	14
24	12	18	16
30	18	17	24
36	26	40	33
42	36	54	47
48	46	69	60

INSTALLING THE FAN

Once you have determined the fan size for your application, look at several different types to determine which will best suit your house. Some types of attic fans do not require any anchors or base: they rest on attic floors or joists and are cushioned with rubber mounts to absorb vibration and seal the air inlet. Flanges on the rim of their shutter assembly cover raw edges, eliminating finish carpentry and painting.

Follow the instructions furnished by the manufacturer of your particular fan. Here are the procedures recommended for one type:

1. Carefully select the best location for the fan. The ideal spot is in the center of the house such as a central hallway.
2. Drill a pilot hole through the ceiling from which you can make exact measurements on both sides of the ceiling.
3. Using a square and a marker, mark out the exact location of the shutter.
4. Score the plaster with a chisel or utility knife along the lines just marked and then cut out the opening with a sabre or keyhole saw.

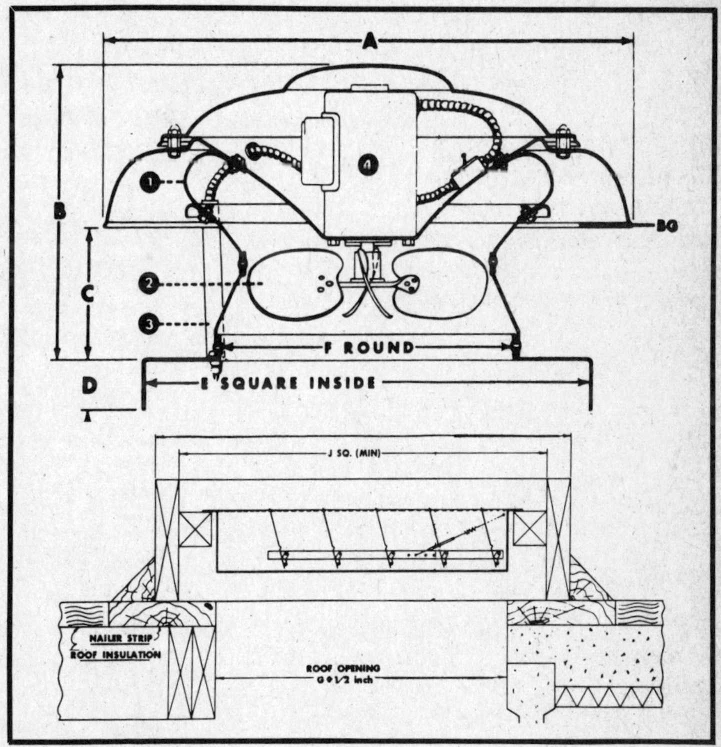

Fig. 12-6. A roof exhaust fan suitable for ventilating flat or shed roofs; note built-up roof curb.

5. Joists may need to be cut and headers installed. But make certain that the building's structure is not endangered. If there's any doubt, call in a consultant.
6. Install the automatic shutter and test its spring tension.
7. Place the fan over the opening and secure it according to the manufacturer's instructions.
8. The fan is wired through the junction box usually attached to the fan housing. Some areas require that a safety switch with a manual reset be installed in the circuit. In case of fire, the fusible link in the switch automatically shuts off power to the fan.

Fan Controls

There are several other ways in which attic ventilating fans may be controlled. The simplest is a conventional wall switch of the type used to control lighting. A special wall-mounted time switch is also available which will automatically shut off the fan at a preselected time. Most have time ranges up to 20 hours. Or the fan may be controlled with a conventional line-voltage thermostat set to come on at the desired temperature.

A more sophisticated control system consists of dual series-connected thermostats, one located outside and one inside. When inside temperatures reach 75°F, the inside thermostat calls for the fan to come on, but if the outside air temperature is more than, say, 80°F, the fan pulling in outside air would only warm the house more. So the contacts on the outside thermostat remain open, allowing no current to pass, if the outside air is above 80°F. At night, when the outside air cools, the contacts close and the fan comes on.

The wiring of the above control is not as complicated as it may seem. All that is required are two line-voltage thermostats wired according to Fig. 12-7.

If your home does have a central air conditioning system, you can still utilize an automatic exhaust fan—as mentioned previously—to lower your attic temperature, lessening the load on the air-conditioning system. The fan must ventilate the entire attic.

Therefore, seal off all openings close to the fan and locate all inlets at the end of the attic opposite to the fan. Wide mesh screen over these openings avoids clogging and restriction.

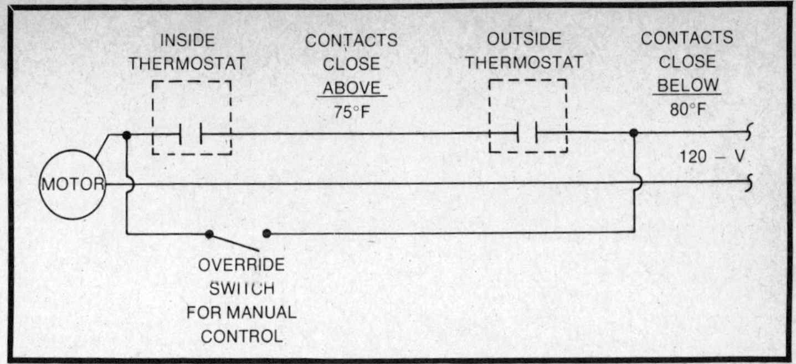

Fig. 12-7. Wiring diagram used on an electric circuit feeding a ventilating fan to provide fully automatic control.

This type of attic fan should be fully automatic so that it will start up when attic heat starts to build up. Most thermostatic controls start the fan when the attic temperature reaches 95°F and shut it off at 90°F. Automatic louvers attached to the fan housing seal the opening when the fan is not operating.

The method for sizing such fans is similar to the method described for sizing a fan to cool the living area of the home; that is, find the cubic feet of the area and select a fan which will move the correct amount of air. In attic spaces, however, one complete air change every five minutes is sufficient rather than one change every minute or every 1 1/2 minutes as suggested for living areas.

Chapter 13
Installing Radiant Heating Cable

Radiant heat is acknowledged to be one of the greatest advances in structural heating since the Franklin stove, and thousands of homeowners all over the country have chosen this type of heat for their homes.

The enormous heating surface of radiant cable means no part of the system need be raised to a high temperature. Rather, gentle warmth radiates downward (or upward) from the surfaces, heating the entire room area evenly, leaving no cold spots or drafts. As with other types of electric heat, there is little worry for the homeowner—no ashes to haul, no fuel to order. All you do is set the thermostats.

There is virtually no maintenance with a radiant heating system as there are no moving parts: nothing to get clogged up, nothing to clean, oil, or grease, and seldom does the equipment wear out! The installation of this system is not out of reach of most do-it-yourselfers; the most difficult part of the entire project being the layout of the system.

A good opportunity to install electric heating cable is during the remodeling of an area within an existing residence—perhaps one where the ceiling plaster is beginning to crack and you're going to cover it with drywall. Or perhaps your basement floor needs repairing and you plan to pour an additional three inches of concrete over your existing floor. This would be the time to install radiant heating cable in the concrete slab.

INSTALLATION IN PLASTER CEILINGS

In order to determine the spacing of the cable on a given ceiling area, take one foot from the room length and one foot from the room width and multiply this new length by the new width. This will give you the usable ceiling area in square feet.

Determine the length of the heating cable you're going to use. Add the reduced room length to the reduced room width and subtract this figure from the cable length. This procedure compensates for the initial run of cable and the U-shaped loops that link each run. Divide this reduced cable length into the usable ceiling area to get the spacing between runs.

For example, let's assume our room measures 14 feet by 12 feet, and has a calculated heat loss of 2000 watts. Subtracting a foot from each dimension gives 13 feet by 11 feet; multiplication gives 143 square feet. From Table 1 we find that a 2000-watt cable is 728 feet long. Subtracting reduced room width and length from cable length (728 minus 24) gives 704 feet. 704 feet divided into the usable ceiling area (143 square feet) gives the spacing between cable runs: 0.203 foot or 2.44 inches.

Figure 13-1 shows a typical heating cable installation as suggested by Tennessee Plastics Inc. for their Ra-Heat cable. However, nearly every brand of heating cable will be installed in exactly the same way, and the procedures are as follows:

1. Nail the outlet box on an inside wall approximately 5 feet above the finished floor (for your thermostat).
2. Drill two holes in the wall plate above this junction box location.
3. Drill two holes through the ceiling lath above the thermostat location.
4. Put the spool of heat cable on a nail, screwdriver or any type of shaft (for unwinding the cable from the spool).
5. Cover the accessible end of the eight-foot nonheating lead wire with non-metallic loom. The loom should be long enough so at least two inches will go on the ceiling surface and reach to the thermostat box, but leave six inches of lead wire inside the box for viewing the identification tags. Never, for any reason, remove these tags. Also, do not cut or shorten the nonheating leads.

Table 13-1. Table of Radiant Heating Cable Characteristics.

CATALOG NUMBER	CAPACITY WATTS	VOLTS	COLOR OF NON-HEATING LEAD	COLOR OF HEATING ELEMENT	APPROX. WT. LBS.	STAND. PACK	LENGTH IN FEET
C2-2 1/2	250	240	Red	Peach Dots	1	6	91
C2-4	400	240	Red	Peach Number Peach Stripe	2	6	151
C2-6	600	240	Red	Purple Number Purple Stripe	2	6	218
C2-8	800	240	Red	Black Number	3	6	291
C2-10	1000	240	Red	Black Dots	4	6	364
C2-12	1200	240	Red	Green Number	4	6	437
C2-14	1400	240	Red	Orange Number Orange Stripe	5	6	509
C2-16	1600	240	Red	Brown Number	5	6	582
C2-18	1800	240	Red	Green Number Green Stripe	6	5	655
C2-20	2000	240	Red	Green Dots	6	5	728
C2-225	2250	240	Red	Yellow Dots	7	3	819
C2-25	2500	240	Red	Orange Number	9	3	936
C2-275	2750	240	Red	Purple Dots	10	3	1028
C2-30	3000	240	Red	Brown Stripe	10	3	1090
C2-33	3300	240	Red	Black Number Black Stripe	10 1/2	3	1200
C2-36	3600	240	Red	Yellow Number	12	3	1310
C2-40	4000	240	Red	Purple Number	14	2	1455
C2-44	440	240	Red	Yellow Number Yellow Stripe	16	2	1600
C2-48	4800	240	Red	Peach Number	17	2	1747
C2-50	5000	240	Red	Brown Number Brown Dots	18	2	1819
C1-1 1/4	125	120	Yellow	Peach Dots	1/2	6	2246
C1-2	200	120	Yellow	Peach Number Peach Stripe	3/4	6	76
C1-3	300	120	Yellow	Purple Number Purple Stripe	1	6	109
C1-4	400	120	Yellow	Black Number	2	6	146
C1-5	500	120	Yellow	Black Dots	2	6	182
C1-6	600	120	Yellow	Green Number	2	6	219
C1-7	700	120	Yellow	Orange Number Orange Stripe	2 1/2	6	255
C1-8	800	120	Yellow	Brown Number	3	6	291
C1-9	900	120	Yellow	Green Number Green Stripe	3 1/2	6	328
C1-10	1000	120	Yellow	Green Dots	3 1/2	6	364
C1-112	1125	120	Yellow	Yellow Dots	4	6	410
C1-125	1250	120	Yellow	Orange Number	4 1/2	6	468
C1-137	1375	120	Yellow	Purple Dots	5	6	514
C1-15	1500	120	Yellow	Brown Stripe	5 1/2	6	545
C1-165	1650	120	Yellow	Black Number Black Stripe	6 1/2	5	600
C1-18	1800	120	Yellow	Yellow Number	7	5	655
C1-20	2000	120	Yellow	Purple Number	7	5	728
C1-22	2200	120	Yellow	Yellow Stripe	8	3	800

6. Run the accessible end (loom and all) of your cable through one of the holes in the ceiling, down the wall, through the plate, and into the thermostat box.
7. Pull the slack out of your nonheating leads and staple them securely to the ceiling. Any excess nonheating lead should be covered with plaster—the same for excess cable. Do not staple or bend the cable.
8. Your next step is to mark the ceiling. First mark a line all the way around the room six inches away from each wall. (This six-inch margin will not be covered with cable and is the reason you deducted one foot from the room width and length. A chalk line is best for marking this. Now notch a yardstick or other straightedge precisely for your calculated spacing. Using the measuring stick, mark the spacing interval on both end margin lines and, if you wish, at points between the end margins (to keep the runs parallel along their entire length). The first cable run should be placed right on one of the margin lines. Keep the

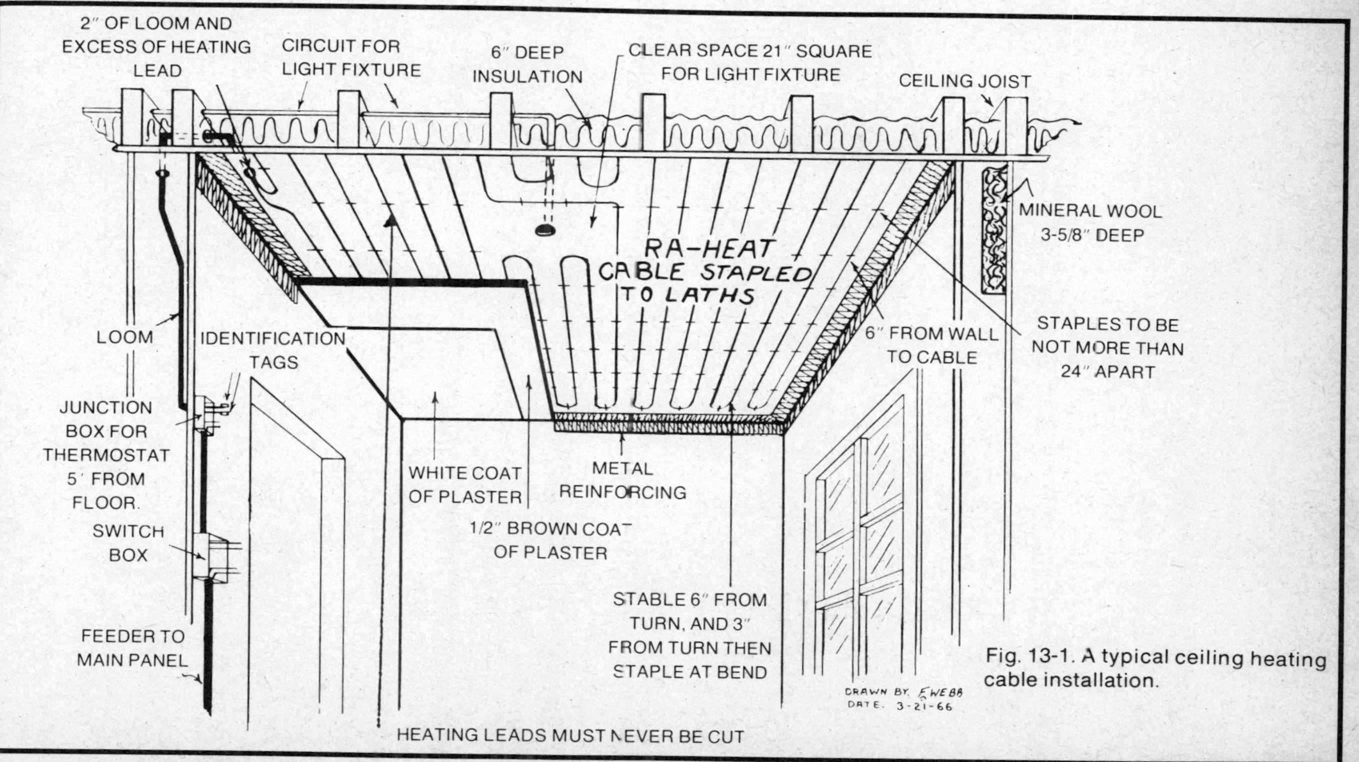

Fig. 13-1. A typical ceiling heating cable installation.

cable at least two inches from metal corner lath or other metal reinforcing.

9. When you are down to the return lead wire you must be back to the starting wall. Cover this lead with the same length of loom as you did the starting lead. Staple this return lead securely to the ceiling and run it through the other hole in the ceiling, down the wall, and into the thermostat box.
10. Take other precautions as shown in Fig. 13-1 and then connect the thermostat, which should be fed by a circuit sized according to the current in amperes drawn by the heating cable. To find the amperes drawn by a cable, divide the wattage by the rated voltage. Your answer will be the load in amperes. Then use the following table to size your wire.

AMPS	TYPE TW WIRE
15 or lower	10 AWG
15–20	8 AWG
20–30	6 AWG
30–50	4 AWG

Always make certain that the heating cable is connected to the proper voltage. A 120-volt cable connected to a 240-volt circuit will melt the cable, while a 240-volt heating cable connected to 120-volts will produce only 25% of the rated wattage of the cable.

INSTALLING CONCRETE CABLE

The heat loss is calculated in the same manner as for any other area (see Chapter 2). For best results, the heating cable should never be spaced less than 2 1/2-inches apart in concrete floors except around outside walls, which may be spaced on 1 1/2-inch (minimum) centers for the first two feet from the wall. The concrete thickness above the cable should be from 1/2 to 1 inch.

The spacing of the cable is found exactly as for ceiling cable, and the installation procedure is as follows:

1. Secure a junction box on an inside wall approximately 5 feet from the finished floor to house the thermostat.
2. Install a piece of rigid conduit from the junction box to the floor to house the nonheating leads.

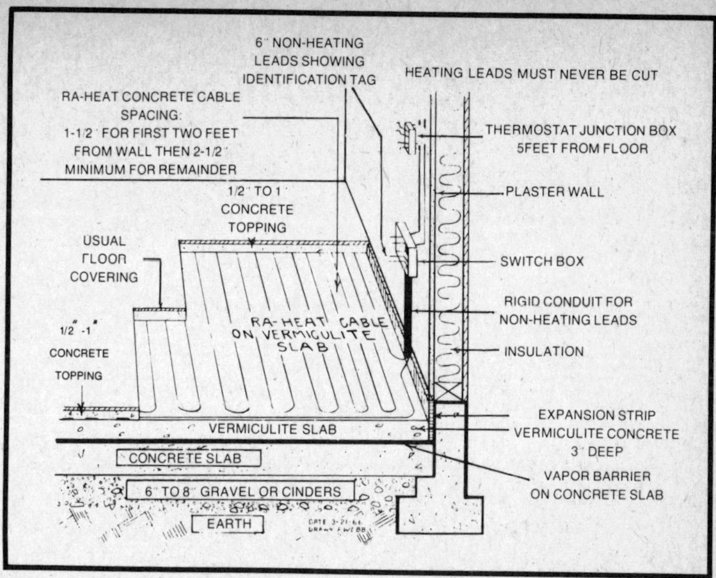

Fig. 13-2. A typical installation utilizing heating cable in a concrete slab.

3. Approximately six inches of the lower end of the conduit should be embedded in the concrete. Install a smooth porcelain bushing on this end to protect the nonheating leads where they leave the conduit. See Fig. 13-2 for an overall picture of a floor-cable installation.
4. Place the spool of heating cable on a nail, screwdriver, or other shaft for unwinding the cable from the spool.
5. Run the accessible end of the eight-foot nonheating lead through the conduit to the thermostat junction box, leaving six inches extending out of the box. Never remove the identification tags or shorten the nonheating leads. If there's any excess nonheating lead, it should be embedded in the concrete, as should excess heating cable.
6. Run the cable along the floor, six inches out from the wall, to the outside or exposed wall, fastening the cable to the floor with staples or masking tape.
7. The cable is usually spaced 1 1/2 inches apart for the first two feet next to the exposed wall, and never less than 2 1/2 inches apart for the remaining area.
8. Run the return nonheating lead wire through the conduit to the thermostat junction box in the same

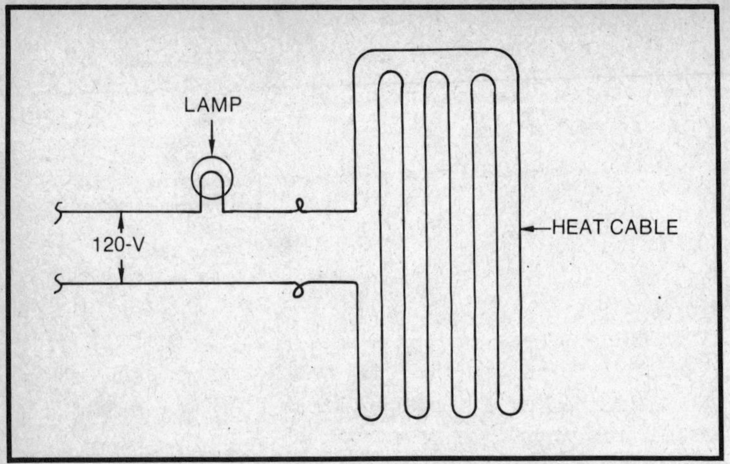

Fig. 13-3. Method of using a lamp socket and 100-watt bulb to test heating cable as it is being installed.

manner as the starting nonheating lead was run in item 5 above.

In general, the concrete slab should be prepared by applying a vapor barrier over 4 to 6 inches of gravel. Then pour 4 inches of vermiculite or other insulating concrete over the gravel, after the outside edge of the slab has been insulated in accordance with good building practice. The heating cable is then installed as described previously.

The homeowner should inspect and test the cable before the final layer of concrete is installed, because once the concrete is poured, it's an expensive matter to repair. First, visually inspect the cable for possible damage during application. Then, with a suitable ohmmeter, check for continuity and capacity of the cable. Concealed breaks may be found by leaving an ohmmeter connected to the cable, and then brushing the cable lightly with the bristles of a broom. Any movement of the meter dial will indicate a fault.

During the pouring of the final coat of concrete it is recommended that the ohmmeter be left connected to the heating cable leads to detect any possible damage to the cable during pouring. Use ordinary concrete for the finish layer, not insulating concrete. If an ohmmeter is not available, a 100-watt lamp may be connected in series with the cable as shown in Fig. 13-3 to detect damage. The lamp will glow as long as the circuit is complete. However, if a break does occur, the lamp will go out and the break can be repaired before the concrete hardens.

Repairs to a broken cable are made by stripping the ends of the cable and rejoining them with a No. 14 AWG pressure-type connector approved for this purpose. The splice must then be insulated with thermoplastic tape to a thickness equal to the insulation of the cable. Use any thermoplastic tape listed by the Underwriters' Laboratories as suitable for a temperature of 176°F.

Once the finish layer of concrete sets, asphalt tile, linoleum tile or linoleum can be laid on the concrete in the normal manner.

Besides the heating of one's home, electric heating cable has many other uses for home owners. It can be embedded in concrete or asphalt surfaces for the removal of ice and snow; protect water pipes exposed to cold weather from freezing; de-ice roofs and gutters; and heat soil in your hotbed or window box. Heating cable is relatively inexpensive and the installation goes fast—just right for the do-it-yourselfer.

SNOW-MELTING WITH ELECTRIC WIRES

The primary consideration for an electric snow-melting system is the amount of heat energy required to melt the snow as it falls on a surface. In an embedded system, the heating elements may consist of a length of copper or alloy wire providing resistance to the flow of electricity, which in turn produces heat.

These elements are either solid-strand conductors or conductors spirally wrapped around a non-conducting fibrous core such as synthetic fiber, glass fiber, or asbestos. Both types are covered with an exterior layer of insulation consisting of silicone polyvinyl chloride, magnesium oxide, or a simliar substance.

The capacity of heating cable to melt snow is proportional to its wattage. Table 13-2 lists the watts per square foot required in a number of states for three classes of systems. Refer to the table, which includes definitions of Class I, II, and III systems, to determine the density of cable for your system.

A plan view of a typical installation is shown in Fig. 13-4. For total snow removal in an area ("A"), select mats to cover the entire area. For large areas it may be desirable to melt snow from only the most frequently used portions such as walks and tracks for autos (as area "B" of Fig. 13-4). Also

Table 13-2. Watts Per Square Foot Required for Class I, II, and III Snow-Melting Systems.

(Common design heat density installed in slab, watts per sq ft of heated area, based on 1966 survey of actual practice as reported by 66 electric utilities)

Class I, Residential; Class II Commercial-Industrial; Class III, Critical*

Location	Class* I	Class* II	Class* III
	Watts per sq ft		
Arkansas			
Ft. Smith	20	40	45
Little Rock	20	30	50
Colorado			
Denver	42	50	60
Pueblo	—	45	60
Connecticut			
Hartford	30	50	70
Middletown	40-60	40-60	60-70
New Haven	40	40	60
Delaware			
Wilmington	30	40	50
District of Columbia	30-40	40-55	55-60
Illinois			
Chicago	40	50	60
Peoria	40	45-50	55-60
Maryland			
Baltimore	30-45	45-55	50-70
Massachusetts			
Boston	40-45	50-60	60-75
Fall River	40	40	60
Springfield	40	40	80
Michigan			
Detroit	40-60	60	60
Jackson	40	60	80
Minnesota			
Minn.-St. Paul	42-75	60-75	70-75
Missouri			
Kansas City	42	40-50	60-70
St. Joseph	42	42	—
St. Louis	40-60	40-60	60
Nebraska			
Lincoln	40-50	40-50	60
Omaha	40-45	60	60
Ironton	30	40	—
Lima	40	40	70
Portsmouth	30-40	40	—
Steubenville	40	45	—
Oklahoma			
Oklahoma City	25	40	45
Tulsa	20	30	40
Oregon			
Portland	42	42	60
Pennsylvania			
Allentown	40	44-55	60
Johnstown	30	40	70
Philadelphia	40	40-60	40-100
Pittsburgh	30-40	30-60	60
Rhode Island			
Providence	45	65	—
Tennessee			
Kingsport	—	42	—
Nashville	40	40	60

Rockford	42	40-60	—
Springfield	40	45-50	55-60
Indiana			
Elkhart	—	42	—
Hartford City	—	35	—
Indianapolis	40	40	40-60
South Bend	—	52	50-55
Iowa			
Dubuque	40	40-60	—
Kansas			
Kansas City	40	50	60
Topeka	40	40	60
Wichita	50	50	50
Kentucky			
Ashland	—	42	—
Maine			
Bangor	40	40	60
Portland	40	40	60
New Hampshire			
Concord	50	50	75
Manchester	—	—	—
New Jersey			
Atlantic City	30	40	60
Morristown	40	50	60
New York			
Buffalo	—	60	—
New York City	35-50	40-50	50-60
Poughkeepsie	40	70	100
Syracuse	40-60	60	60
North Carolina			
Charlotte	42	30-42	42
Ohio			
Canton	30	36	—
Cincinnati	40	50	60
Cleveland	40	—	45-55
Columbus	30	40	50
Findlay	—	40	60
Vermont			
Bennington	50	50	75
Burlington	50	50	75
Virginia			
Richmond	30	40	—
Abingdon	30	30	—
Washington			
Spokane	30-40	30-45	—
West Virginia			
Wheeling	35	35-50	—
Bluefield	—	40	80
Parkersburg	30	45	60
Beckley	40	40	—
Charleston	40	45	50
Morgantown	30	45	60
Wisconsin			
Milwaukee	45	50-65	65

*Class descriptions:

I, Residential: Residential walks or driveways and interplant areaways—will allow snow to completely cover area temporarily, but will not accumulate for worst conditions of 98% of snow frequency. Minimum recommended installed intensity 30 watts per sq ft.

II, Commercial-Industrial: Commercial sidewalks, steps and driveways—will allow snow to completely cover area temporarily, but will not accumulate for worst conditions of 100% of snow frequency. Minimum recommended installed intensity 40 watts per sq ft.

III, Critical: Toll plazas of highways and bridges, and aprons and loading areas of airports—will melt snow immediately for 98% of snow frequency. Minimum recommended intensity, 60 watts per sqft. Per Article 422-7(b) of the National Electric Code, installed heating intensity of embedded cable or wire systems shall not exceed 120 watts per sq ft.

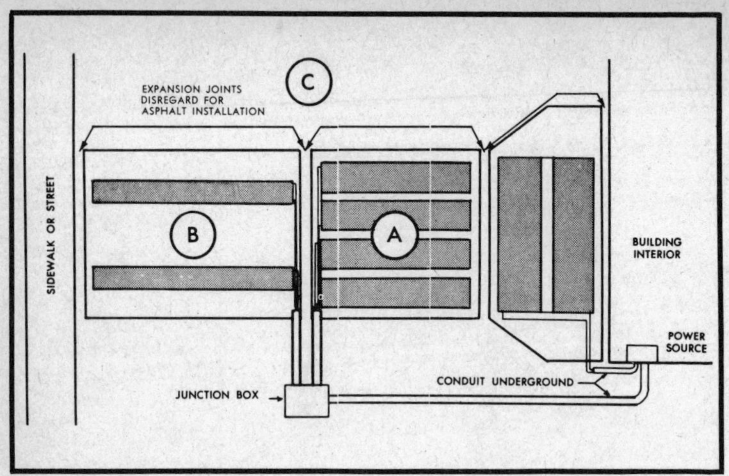

Fig. 13-4. Plan view of a typical snow-melting application.

consider the possibility of feeding each mat with a separate circuit so that areas within the system can be heated independently.

You will have to locate expansion joints in all concrete surfaces (a) whenever the slab changes size or direction and (b) spaces no more than twenty feet apart (see "C" of Fig. 13-6). Since the mats should never be run through expansion joints, select units of the largest size which will fit *between* these joints. The junction boxes should also be located so that the maximum number of heating mats can be accommodated by each box. See Figs. 13-5 through 13-8.

Electrical wiring must conform to requirements of the National Electrical Code and all junction boxes should be of a

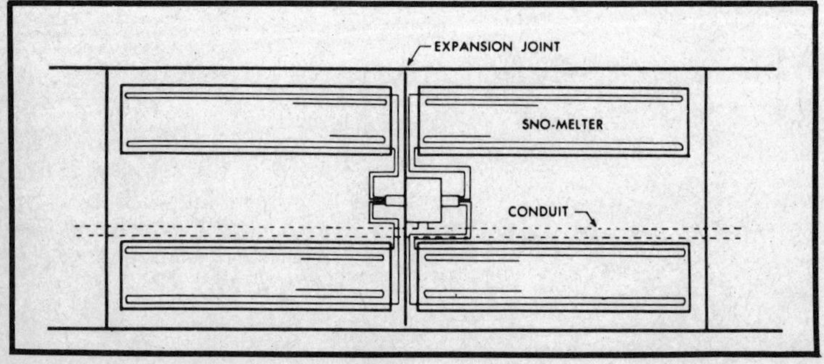

Fig. 13-5. Plan view of an installation showing one junction box being used to connect four mats.

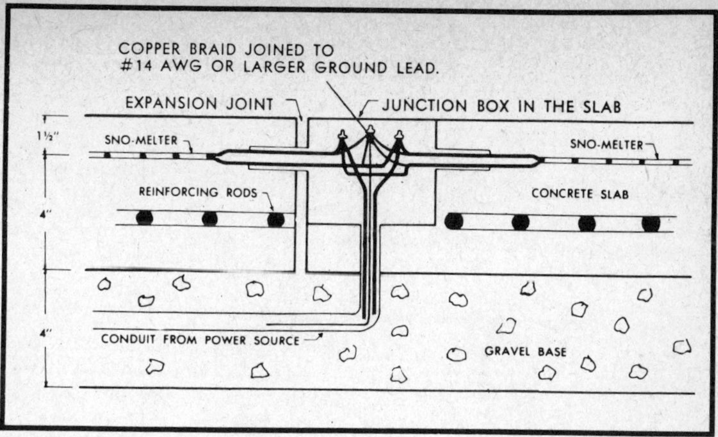

Fig. 13-6. Method of connecting electric feeder to leads of snow-melting mats.

type approved for outdoor use. Locate the junction boxes adjacent to the concrete slab if possible. However, they may be mounted in the slab if this is necessary, provided cast-aluminum or iron boxes are used.

Overcurrent protection of the proper size must be used on all circuits feeding the heat cable. Double-pole, single-throw switches, or two-pole circuit breakers—to open both sides of the line—should be used. Locate these protective devices in the main electric panel or in any protected remote location, but include a pilot lamp on the load side of the circuit to indicate

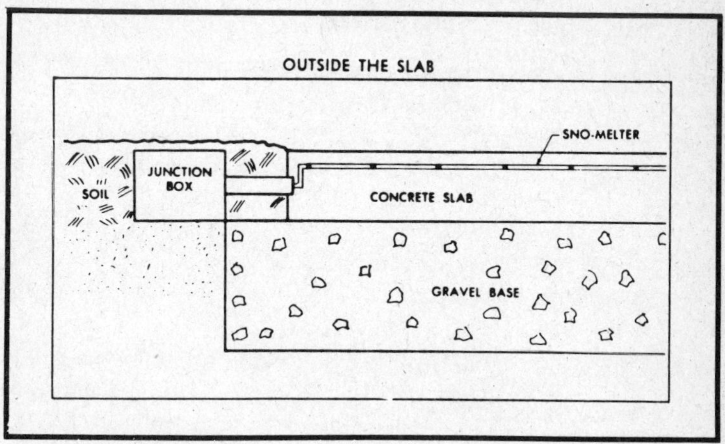

Fig. 13-7. Section through a heated concrete slab showing junction box, pipe sleeve, heating cable, gravel base, and concrete slab.

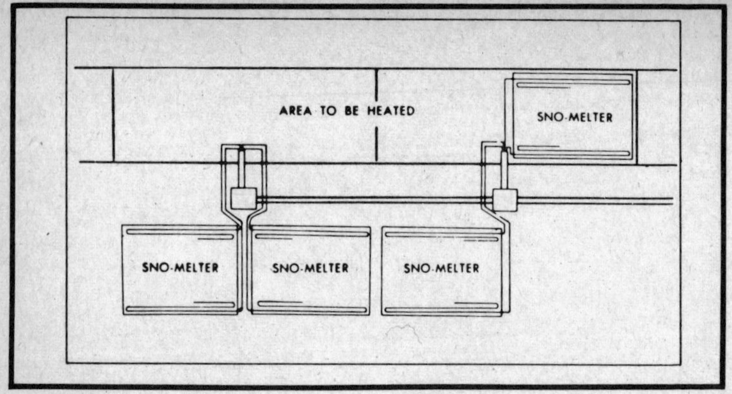

Fig. 13-8. Plan view showing location of electric junction boxes and the heating mats connected.

when the heat cable is energized. An automatic heat cable switch is also available and is described later in this chapter.

The concrete slab in which the heat cable is embedded should be well-planned before installing the cable, and the following should be kept in mind: (1) a well-prepared base; (2) adequate drainage to prevent accumulation of water and resultant heaving (frost damage); (3) sufficient slab thickness (four-inch minimum reinforced with crushed-rock aggregate); (4) suitable reinforcements; (5) sufficient number of expansion joints.

With these basic suggestions in mind, you are now ready to begin the installation which will consist of the following steps:

1. Locate junction boxes according to the suggestions given earlier and as planned on your layout drawing. Before placement of the reinforcing mesh, run the power supply conduit underground, outside the slab or in a prepared base as previously described.
2. Check for continuity of the mats from electrical leads to ground with an ohmmeter on its highest range (to assure that the mats have not been damaged during shipment); this reading should be infinite.
3. Thread the mat lead wires thorugh the conduit stubs and bushings into the closest junction box and then lay mats in place to check the layout and spacing. Leave sufficient slack in the lead wires to permit handling the units. Lay the mats temporarily out of the way for the initial pour of concrete.

4. The concrete can now be poured to within 1 1/2 inches of the finished grade level. Then roughly level the concrete off. As an area is being poured, reposition the heating mats with the heater wire face down and mesh side face up. Bury all excess wire in the concrete.
5. Again check all units with an ohmmeter to ensure that no damage has occurred during the installation. Then continue by covering the heating units with the final 1 1/2 inches of concrete.
6. Proceed with the pouring of each successive area to within 1 1/2 inches of the final level, position the units, check for continuity, and finally pour the balance of the concrete and finish the surface. It is recommended that each slab area within an expansion joint be poured and finished individually.
7. As the concrete sets up and can be walked on, make up all wire splices at the junction boxes and the feeder circuits installed earlier. Twist together a length of the copper grounding braid from all leads and positively connect to a continuous, No. 14 AWG or larger, insulated copper wire extending to the distribution panel ground. All wiring should be done in compliance with the National Electrical Code.
8. Secure all splices with approved pressure-crimped connectors or set-screw-type wiring clamps. Thoroughly tape all power splices with plastic electrical tape to make them waterproof. To further protect the connections, place them inside the junction box and fill the box with wax or electrical potting compound.

Before we get into the control of snow-melting cable, there are a few precautions one should take when installing the system. To begin, never pour concrete in freezing weather as the final result will be of poor quality, necessitating early replacement of the concrete (which means the heating mats will also be lost). You should never walk on the units during the pouring of the concrete and they should never be struck with a shovel or other tool. Don't turn on the electricity until the concrete has thoroughly cured.

CONTROL OF HEATING CABLE

There are several ways in which the heating cable can be controlled. The simplest, of course, is to use a single- or

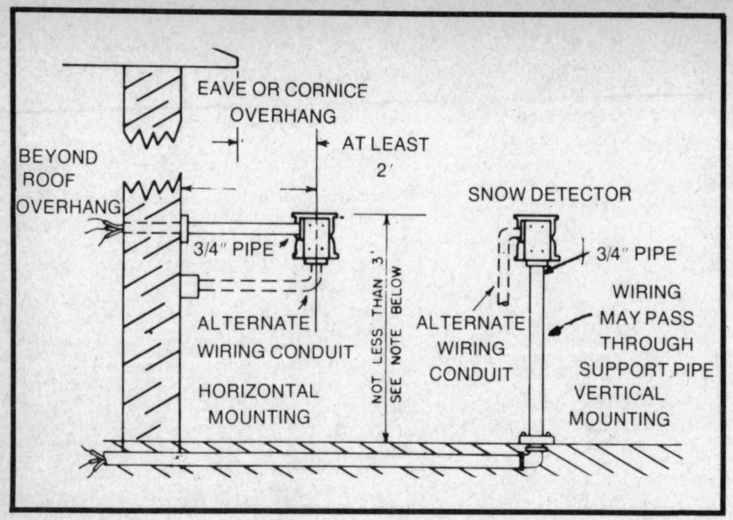

Fig. 13-9. Methods of mounting automatic snow detector switch.

double-pole wall switch and merely turn the switch on when it begins to snow. It may also be controlled in a similar manner at the circuit breaker in the panelboard. The most sophisticated installations include an automatic switch.

Automatic switches specifically designed for detecting snowfall are available from the same manufacturers that supply the cable (See Appendix I). This type of switch immediately closes a low-voltage circuit to actuate the snow-melting equipment. It should be mounted to collect representative snowfall near the surface to be melted.

The operating principle of an automatic switch is really very simple. The detector is mounted with the snow-collecting funnel pointing up as shown in Fig. 13-9. Electric eight-watt heaters are attached to a bimetallic "bridge" unit which is ambient-compensated. Therefore, any thermal unbalance of this bridge actuates a precision snap switch, which in turn closes a control circuit. If the ambient temperature is above 35°F, a built-in thermostat prevents an output melting signal from occuring. However, at 35°F or lower, the thermostat closes and activates the unit.

The surface of the collecting funnel is kept above 32°F automatically, so when the first snowflakes fall into the funnel, they are converted to water drops which pass through the funnel opening into a small collector cup attached mechanically and thermally to the bimetallic bridge. Enough snowflakes to create a few drops of water will unbalance the bridge and start the melting operation. When snowfall ceases, the water in the cup keeps the bimetal bridge thermally unbalanced, and the sensor switch remains closed. However, this water (when not replaced by more snowfall) slowly evaporates from the heat. When the cup is dry, the bridge returns to balance and the sensor switch opens the circuit, stopping the melting process.

Installation Instructions

Select a mounting position where the switch will not be sheltered from snowfall by roof overhang, trees, shrubbery, etc. Avoid locations where the snow is diverted to horizontal or upward motion by wind movement over the roof and similar surfaces. The funnel should point upward where it can collect falling snowflakes but not water dripping from eaves or other overhanging projections. The unit should be no less than three feet above the ground and may be supported on 3/4-inch iron pipe as shown in Fig. 13-9.

You may want to mount the switch close to the ground if the detection of drifting snow is desired. However, low mounting offers the possibility of accidental or mischievous pollution or damage by humans or pets. A height of five to eight feet above the ground is generally practical, and minimizes the possibilities for damage and pollution.

As mentioned previously, the wiring is for 24 volts, but should be enclosed in a suitable metal or plastic conduit. Splice and solder the connections to the leads in the box, noting color-coding and recommended wiring diagrams as shown in Figs. 13-10 through 13-13. The separate power transformer (120/24 volts) should be mounted indoors or in a weatherproof enclosure.

When the wiring is completed, the operation may be checked as follows. Turn on the power supply. If temperature is above 35°F at the snow-melting switch location, remove the junction box cover, and jump the terminals of the thermostat switch. Wait for approximately five minutes, then drop 5 to 10 drops of water into the funnel. In about one minute the sensor switch should close and turn on the melting system. If the system operates, remove the jumper wire from the thermostat terminals and replace the junction box cover. In doing so, be careful not to damage the copper capillary tube that extends

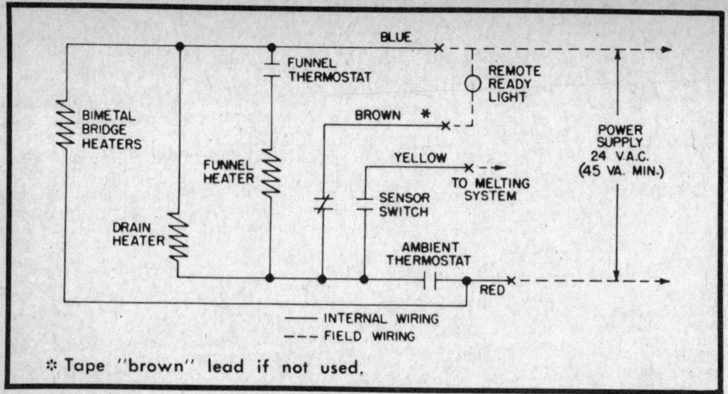

Fig. 13-10. Internal schematic and partial wiring of an automatic snow-melting switch.

under the cover to the plate-type temperature element on the side.

If the temperature at the unit is below 35°F, it is not necessary to jump the thermostat. Merely turn on the power supply, wait about five minutes, and add a few drops of water. The sensor switch should operate as described above.

No servicing is required except occasional inspection of the detector unit to see that the funnel has not been clogged by foreign matter such as leaves, seeds, cinders, etc. Fine dirt particles may build up in the collector cup attached to the bimetal bridge. Ordinarily, an accumulation of dirt in the cup will have little effect on sensitivity or "melting extension time." If the dirt, however, has dried to a solid cake, it may have some effect on the performance of the unit.

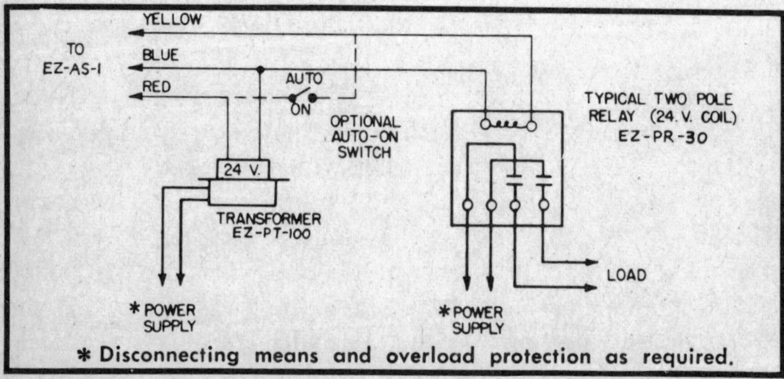

Fig. 13-11. Typical wiring diagram showing connections of a two-pole relay with separate transformer to an automatic switch.

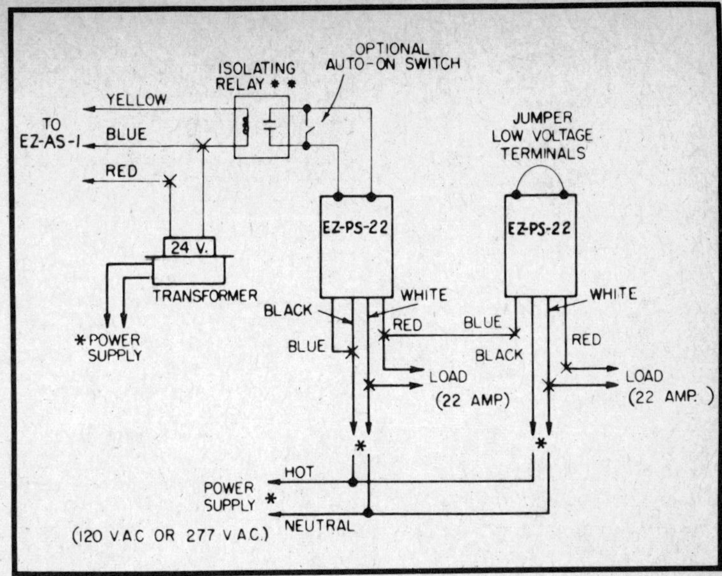

Fig. 13-12. Typical field wiring of an automatic switch to transformer and power switches.

If you find it necessary in your area to extend the melting time beyond the normal operation, an "off-delay timer" may be used. This type of timer keeps the melting equipment in

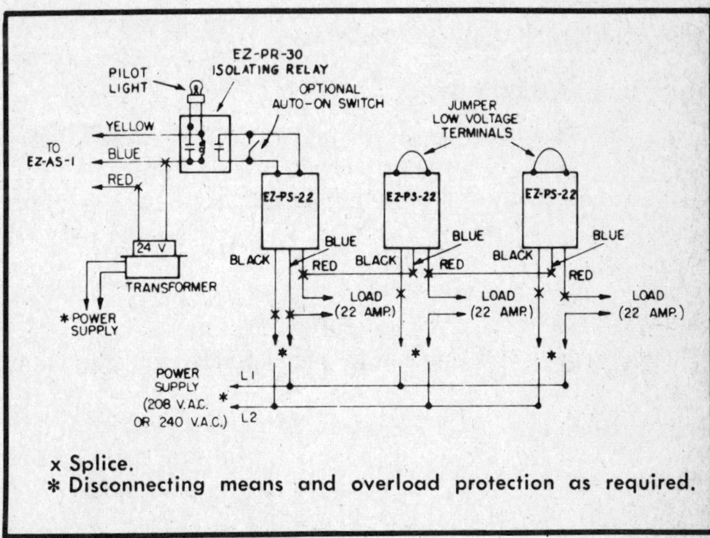

Fig. 13-13. Typical field wiring of an automatic switch to transformer and 120-volt switches.

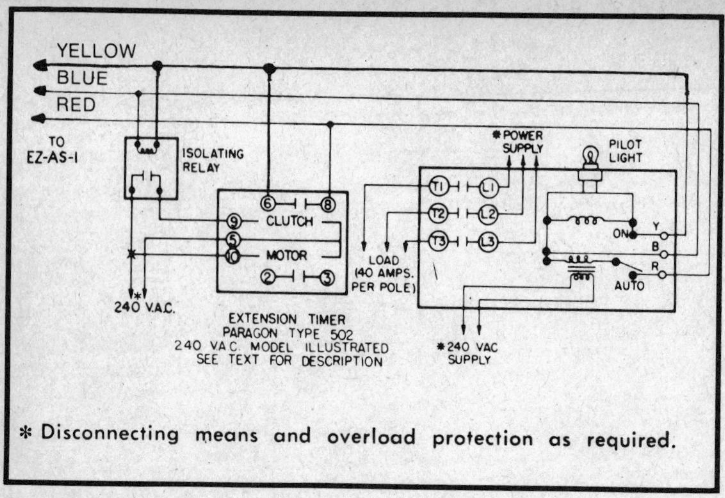

Fig. 13-14. Typical wiring connections to an automatic switch using melting extension timer and a contactor unit.

operation for a preset time after the sensor switch has opened. This extra time is in addition to the nominal 45 minutes after the snow stops falling. If more snowfall is detected during the timer extension period, the timer will reset to again provide the full preset extension or delay time. A typical wiring diagram of a detection unit using a melting extension time is shown in Fig. 13-14 and a pictorial drawing of a typical snow-melting automatic switch is shown in Fig. 13-15.

HEAT CABLE FOR FREEZE PROTECTION

It's sometimes necessary to locate water pipes outdoors where they are subject to freezing in low temperatures. Such applications include mobile home plumbing and livestock tank hydrants.

Automatic and manually-controlled heat tapes are available which are specifically designed for use on outside water pipes. They are manufactured in ratings from 7 to 10 or more watts per foot for use in various applications around the home or farm.

The method used for a pipe-heating installation will be determined largely by the heat requirements and the length of pipe to be heated. Mineral-insulated cable, for example, can be run straight along the pipe as shown in Fig. 13-16 or wrapped as shown in Fig. 13-17. For short lengths of pipe, and where highly concentrated heat is required, it may be necessary to

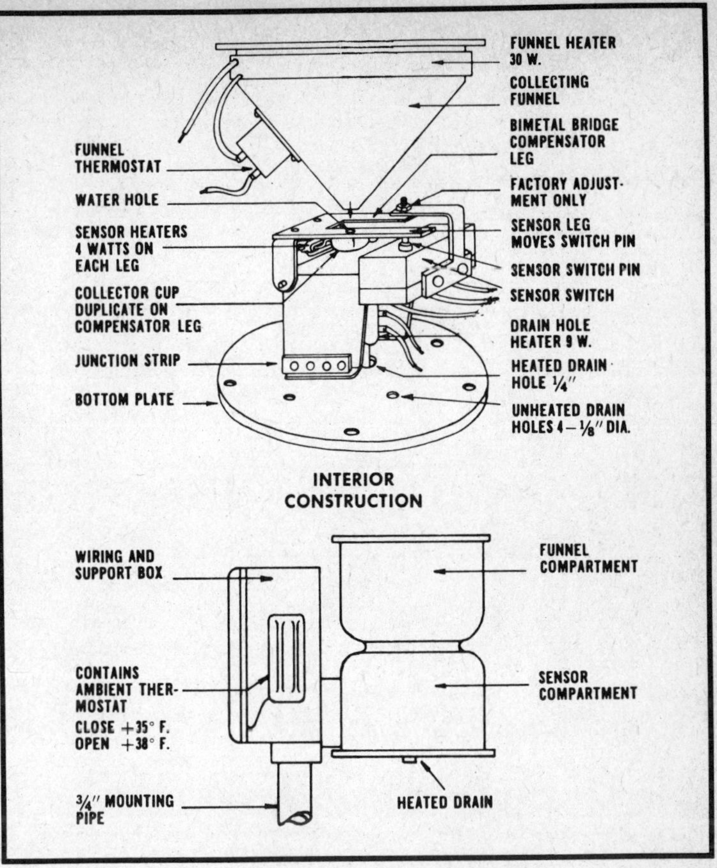

FUNNEL HEATER 30 W.
COLLECTING FUNNEL
BIMETAL BRIDGE COMPENSATOR LEG
FACTORY ADJUSTMENT ONLY
SENSOR LEG MOVES SWITCH PIN
SENSOR SWITCH PIN
SENSOR SWITCH
DRAIN HOLE HEATER 9 W.
HEATED DRAIN HOLE ¼″
UNHEATED DRAIN HOLES 4 – ⅛″ DIA.
FUNNEL THERMOSTAT
WATER HOLE
SENSOR HEATERS 4 WATTS ON EACH LEG
COLLECTOR CUP DUPLICATE ON COMPENSATOR LEG
JUNCTION STRIP
BOTTOM PLATE

INTERIOR CONSTRUCTION

WIRING AND SUPPORT BOX
FUNNEL COMPARTMENT
CONTAINS AMBIENT THERMOSTAT
CLOSE +35° F.
OPEN +38° F.
SENSOR COMPARTMENT
¾″ MOUNTING PIPE
HEATED DRAIN

Fig. 13-15. Typical snow-melting switch.

spiral the heating cable around the pipes; cable length required when spiraled around pipe is shown in Table 13-3.

All wiring to these heating units should be in accordance with local and national electrical codes. Heat tapes designed

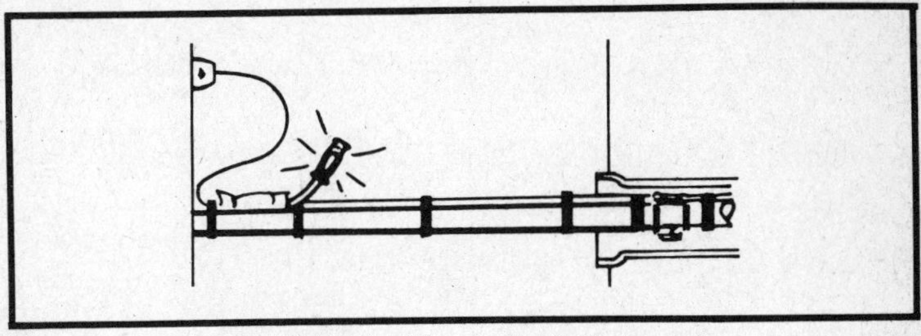

Fig. 13-16. Heat cable run straight along pipe.

Table 13-3. Degree of Protection and Cable Length Required for Common Pipe Sizes. NOTE: Specifications Vary Depending on Manufacturer.

Method of Wrapping	Nominal Pipe Size	3/8″	1/2″	3/4″	1″	1-1/4″	1-1/2″	2″	3″	4″	6′
Straight Along Pipe	Degrees of Protection (F.)	−50°	−40°	−27°	−13°	−3°	0°	+6°	+15°	+20°	+25°
	Length Required per Foot of Pipe	1′	1′	1′	1′	1′	1′	1′	1′	1′	1′
Spiraled Around Pipe 3 turns/ft.	Degrees of Protection (F.)	—	—	−50°	−40°	−33°	−30°	−27°	−20°	−17°	−13°
	Length Required per Foot of Pipe	1′3″	1′4″	1′6″	1′8″	1′10″	2′1″	2′4″	2′11″	3′11″	5
Spiraled Around Pipe 6 turns/ft.	Degrees of Protection (F.)	—	—	—	−50°	−50°	−50°	−50°	−50°	−50°	−50°
	Length Required per Foot of Pipe	1′9″	2′	2′5″	2′8″	3′1″	3′6″	4′2″	5′8″	7′7″	10′3″

Fig. 13-17. Heat cable run in a serpentine pattern.

especially for the homeowner are available with built-in thermostats and a conventional plug for connection to any convenient outlet that will carry the load. The installation instructions for one type of pipe-heating cable follow:

First determine the size of the pipe to be protected and the lowest ambient temperature to which the pipe will be subjected. Then check Table 13-3 to determine the method of wrapping and the length of heat tape required for your particular application. Spirally wrap or apply straight as indicated in the tables for your specified condition. However, be sure not to let the heater wires touch, cross, or overlap. The unit may overheat and burn out if any of these conditions occur. Neither can the unit be made longer or shorter. If it is made shorter it will overheat; if made longer, it will not give off enough heat.

Continue to wrap the unit evenly, spacing each wrap until the frost line is reached; friction tape may be used to hold the end in place while you are wrapping the heat cable. If there is a portin of the tape left unused, do not overlap the wires, but either rewrap the cable—spacing the wires closer together—or let the excess hang loose.

For models with a built-in thermostat, place the thermostat parallel to the pipe and hold it in place with friction tape. When installed in this manner, the thermostat will automatically turn the cable on at 35°F and off at 45°F.

Your next step is to apply black waterproof outerwrap insulation (Fig. 13-18), overlapping halfway with each wrap. Start at the bottom of the pipe and work up so the outer wrap will shed water. Continue, evenly spacing until the pipe is completely covered.

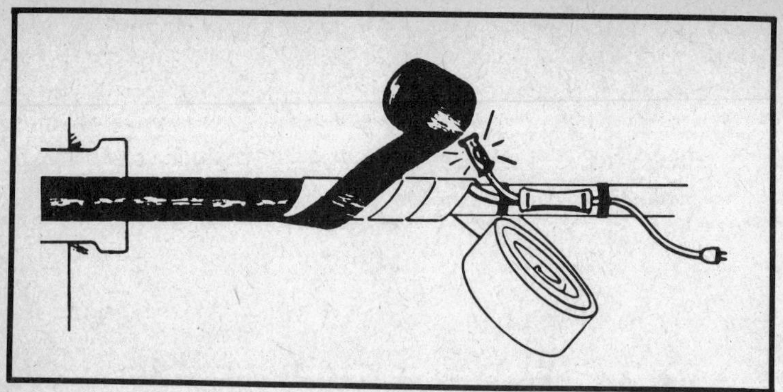

Fig. 13-18. Method of applying black waterproof outerwrap insulation.

SOIL HEATING CABLE

Getting new plants out early is important to both home and commercial gardeners, particularly for food crops such as tomatoes, peppers, eggplant, and other easily transplanted vegetables and flowers. The heat/growth chart (Table 13-4) lists some common vegetables and flowers for which early seed starting is desirable.

Soil heating cable may be installed in your seed bed by following these instructions:

1. Measure the area of your seed bed and multiply by 10 watts per square foot in mild climates and 15 watts per square foot in the colder states.

Table 13-4. Characteristics and Temperature Recommendations for Popular Flower and Vegetable Seed Germination.

	Seed	Recommended Soil Temp F°	F° Air Temp	Approx. Days Germination	Seeding To Cold Frame	Seeding To Field
VEGETABLES	Tomato	68°-75°	70°	3-5	2-3 weeks	5-7 weeks
	Pepper	70°-75°	65°-75°	12-15	3-4 weeks	7-8 weeks
	Cabbage	70°	65°-70° Day 55°-60° Nite	———	1-2 weeks	4-6 weeks
	Broccoli	70°	65°-70° Day 55°-60° Nite	———	1-2 weeks	4-6 weeks
	Onion	70°	———	———	———	5-7 weeks
FLOWERS	Aster	68°-86°	65°-70°	6-12	———	———
	Dahlia	68°-86°	65°-70°	9-16	———	———
	Marigold	68°-86°	65°-70°	7-15	———	———
	Petunia	68°	65°	5-14	———	———
	Zinnia	68°-86°	65°-70°	5-12	———	———

2. Determine cable spacing by dividing the total wattage required by the watts per linear foot of the cable you intend to use. For example, if your location calls for 15 watts per square foot and the area you wish to heat covers 30 square feet, you need 30 times 15 or 450 watts of heating capacity. The following table (most cables' specs are similar) indicates a 420-watt cable—the closest choice—is 160 feet long.

Cable Length	Total Watts (at 120 volts)
6′	21
12′	42
24′	84
36′	126
48′	168
60′	210
80′	280
120′	420
160′	560
200′	700

3. The spacing between the cable and outside walls of the hotbed is usually half the spacing between cable runs. A typical installation is shown in Fig. 13-19.
4. Cover the cable with a two-inch layer of loose soil or sand. Then place a 1/2-inch-mesh screen or hardware cloth on top of the soil or sand: this will prevent damage to the cable when sharp garden tools are used in the hotbed. Notice in Fig. 13-19 that the thermostat is buried in the soil in the same area as the cable to expose it to the same soil temperature. Be careful not to permit the cable to touch the thermostat sheath as this would cause the thermostat to shut off before the soil was sufficiently heated.
5. Lay the cable in position carefully to avoid damaging the insulation sheath. Kinks or twists may damage or break the cable. Never overlap, cross one section of cable over another, or shorten the cable: any of these may cause the cable to burn out.

PREVENT ROOF AND GUTTER DAMAGE

When snow melts, look out for costly and dangerous ice damage. Even in below-freezing weather, snow on the roof will

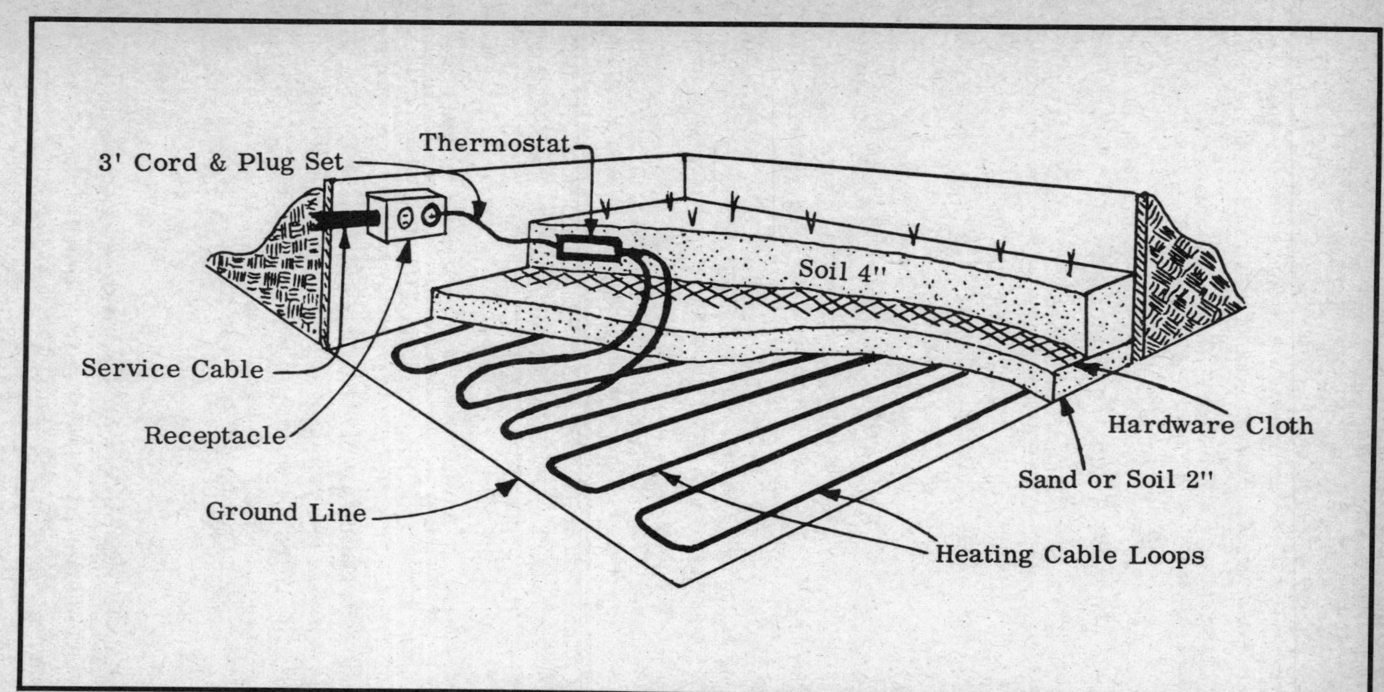

Fig. 13-19. Typical application of heat cable in hot bed.

often melt because of heat from the sun and heat escaping from the house itself. As this melted snow reaches the roof overhang, much of it freezes due to the colder temperature at this point. Once the ice dam has formed, water from subsequent thawings backs up under the shingles and eventually leaks into the house, soaking the plaster, insulation, and interior finishes.

This problem can be preventing by installing electric heating cables in the gutters, downspouts, and along the lower edge of the roof.

Begin by measuring the length of each roof edge, the distance from the roof edge to the inside wall, and the length of the downspout. Then refer to the following table to determine the amount of heat cable required for your home.

Width in Inches	Height in Inches	Cable Length per foot of edge
AVERAGE SNOW FALL		
	12	2.7
18	24	4.2
	36	5.4
HEAVY SNOWFALL		
	12	3.8
14	24	5.6
	36	7.0

To illustrate the use of these tables, assume that the roof edge of a house is 60 feet, while the height (distance from the roof edge up the roof to the inside wall) is 12 inches. We will further assume that this house is located in an average snowfall area requiring the cable to be installed on 18-inch centers; if the house were located in a heavy snowfall area, the cable should be located on 14-inch centers. With this data we can calculate the total length of the cable by:

2.7 (from table) × 60 (length of roof edge) = 162 feet

Adding two 10 foot lengths of downspout gives a total of 182 feet of cable required for the installation.

Most kits for melting ice and snow come complete and ready for installation. Even a supply of aluminum shingle clips is included to hold the cable in position.

As with other types of heating-cable applications, never allow the wires to touch or overlap, and never shorten the heating portion of the cable.

Begin the installation by installing the shingle clips to the underside of the top layer of shingles at the required spacing. Continue by installing the shingle clips at the roof edge in order to separate and clamp the heating cable where it is to dip into the gutter. If the cable is to be installed on a metal, built-up, or other roof without shingles, cut 4-inch squares of shingle and attach them to the shingle clips. Then cement these patches to the roof. For installation on slate roofs, the shingle spurs on the clips can be bent down, epoxy cement applied, and the clip slipped under the slate.

If the outside temperature is below 35°F when you install the cable, it will be stiff and difficult to work. However, if you plug it into a 120-volt outlet for a short time, the wire will become flexible and easier to handle as it warms up.

The cable is strung as shown in Fig. 13-20; that is, the shingle clips are carefully bent over the cable to hold it in place and the roof cable is then attached to the gutter cable by means of clips such as the one shown in Fig. 13-21.

Most manufacturers recommend that the power source (120-volt supply) for the heat cable be within 10 feet of the heating portion of the cable. One way to do this is to install a weatherproof outlet in the soffit or gable end of the house. A better way is to install a circuit and a junction box in the attic space near the end of the cable. Drill a small hole in the soffit;

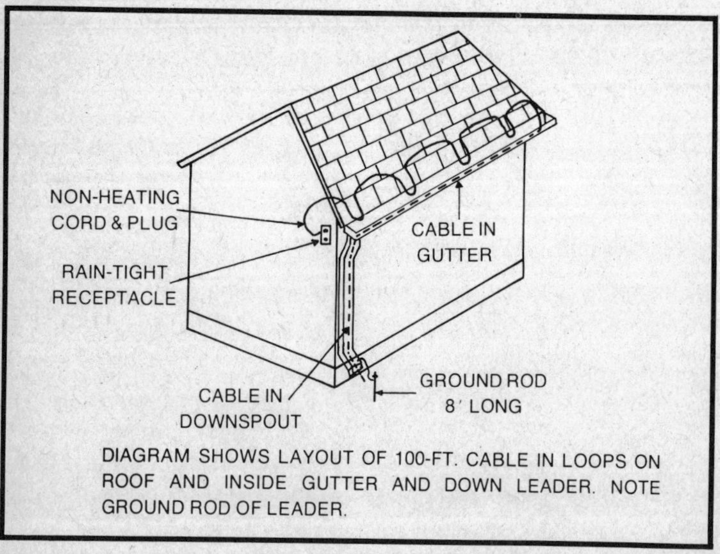

Fig. 13-20. Method of stringing cable.

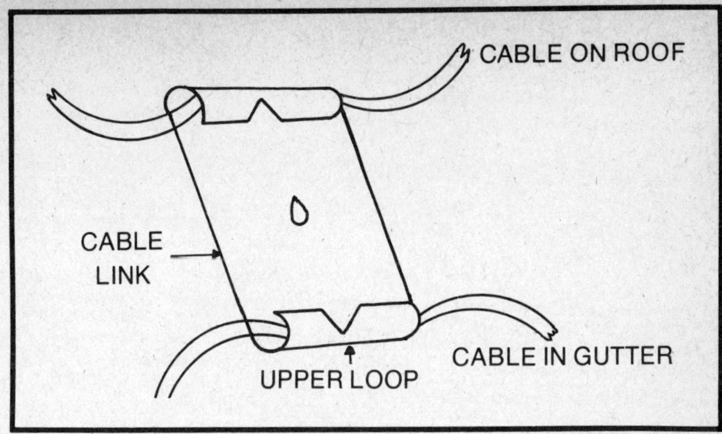

Fig. 13-21. Clip used to attach roof cable to gutter cable.

remove the plug and insert the nonheating portion of the cable through the hole. The area around the hole (through the soffit) should be sealed with patching compound. Run a switch leg from the junction box to a heavy-duty, 20-ampere single-pole switch, located conveniently and wire the cable to the junction box. Then when the cable is needed, all that is necessary is to flip the switch. It's advisable to use a switch with an indicator light, so you'll be less likely to forget the cable is on (which can be expensive).

A 5/8-inch grounding rod, eight feet long, should be driven in the ground at each downspout location as shown in Fig. 13-20. A short piece of No. 12 AWG bare copper wire is then connected to the ground rod by means of a ground clamp and to the downspout with a lug and self-tapping sheet-metal screw. This completes the installation.

Chapter 14

Stretch Your Swimming Season With A Pool Heater

With a swimming-pool heater, you, your family, and your friends can get twice the fun out of your pool—for twice as long. A pool heater to heat the water combined with a few infrared heaters along the edge to take the chill off the air will let you stretch your swimming season from spring to fall, instead of just the few short summer months. Just think of the exhilaration of diving into a warm pool from the cool autumn night air for a cozy moonlight swim.

A swimming pool—at today's prices—is certainly not cheap and a few relatively inexpensive accessories will let you and your family stay in the water longer, helping you get all the enjoyment that's possible out of your investment.

Air at 72°F is comfortable. Water at this temperature, however, chills the average swimmer within an hour because water takes heat from the body much faster than air. For extended swimming, many people prefer a water temperature around 78°F. There are no hard and fast rules governing what constitutes a comfortable water temperature, however. You are the best judge of the temperature at which you are the most comfortable.

SIZING THE POOL HEATER

The size of heater you'll need for your pool depends upon several things:

1. The size of your pool.
2. The climate of the area in which you live.
3. The temperature at which you want to swim.

In order for you to get the most enjoyment and value out of your pool heater, it is important that you select the right size for your circumstances. A heater that is too small will certainly not be able to keep your pool at the temperature you desire, while one that is oversized will eat up fuel and cost you more money initially. Either way, you are wasting money.

On the other hand, a properly selected and installed swimming-pool heater will keep your pool at the desired temperature for season after season.

In general, the size of heater for your pool can be found by following the four steps outlined below:

1. Calculate the pool area in square feet.
2. The surface heat loss included in Fig. 14-1 is based on a 3.5 mph average wind velocity at the water surface. For greater wind velocities, make corrections as follows:
    a. For 5 mph wind velocity, multiply by 1.25.
    b. For 10 mph wind velocity, multiply area by 1.50.
3. Determine temperature differential (between air and pool temperature) as follows:
    a. Determine the pool temperature desired.
    b. Obtain the average temperature during the coolest month you'll want to swim.
    c. Subtract the average temperature from the desired pool temperature to find the temperature differential.
4. Refer to Selection Guide (Fig. 14-1).
    a. Select the differential column that is closest to your temperature differential (from 3.c. above).

POOL HEATER SELECTION GUIDE

TEMPERATURE DIFFERENTIAL BETWEEN AIR and POOL WATER			10°F	15°F	20°F	25°F	30°F
MODEL NO.	INPUT BTU/HR NATURAL	(L.P.)	POOL SURFACE AREA IN SQUARE FEET**				
SPGO-80	80,000	(69,000)	533 (460)	355 (306)	266(230)	213(184)	178(153)
SPGO-160	160,000	(160,000)	1067(1067)	711 (711)	533(533)	426(426)	355(355)
SPGO-240	240,000	(225,000)*	1600(1500)	1067(1000)	800(750)	640(600)	533(500)
SPGO-300	300,000	(240,000)*	2000(1600)	1333(1067)	1000(800)	800(640)	667(533)

*PROPANE ONLY **SURFACE AREA IN PARENTHESIS ( ) IS FOR L.P. GAS

Fig. 14-1. Pool Heater Selection Guide (Courtesy Rudd Manufacturing Co.).

b. Read down this column to an area equal to or slightly greater than the area calculated in step #1 or step #2 (whichever is greater) above.

c. Read to the left to select the appropriate heater.

Now let's review the four basic steps to make certain that you fully understand each one. To find the area for rectangular pools, we obviously multiply the length by the width. For round pools, use the formula $Area = \pi r^2$ or use the following chart.

Sizing Guide For Round Pools

Diameter	Area (Square Feet)
8′	50
10′	78
12′	113
15′	177
18′	254
21′	346
22′	380
24′	452
27′	572
28′	615

Oval pools are a little more difficult to calculate the area of than round or rectangular ones. However, the following chart will help you to determine the approximate area.

Sizing Guide For Oval Pools

Width	Length	Area (Square Feet)
8′	12′	82
10′	15′	128
12′	18′	185
12′	24′	257
15′	25′	327
15′	26′	342
15′	27′	357
15′	30′	402
15′	32′	432
16′	24′	329
16′	32′	457
16′	40′	585
18′	34′	542

With this data covered, we will assume that we want to size a pool heater for an oval pool 15 feet by 30 feet in length. Looking at the chart above, we see that the approximate area is 402 square feet. We have just completed Step 1.

Moving on to Step 2, we have determined through information obtained from a nearby weather station (or a library) that in early May and early October (the months we will probably use the heat the most) that 5-mph winds are typical in our area. So we must multiply our pool area (402 square feet) by the factor 1.25 to obtain an adjusted area of 503 square feet. This will be the figure used in all calculations to follow.

In order to perform Step 3, from our previously mentioned sources we find that the average temperature in our area in early May and early October is 65°F. If we desire to heat the pool water to a temperature of 78°F, we then have a 13° (78° – 65°) temperature differential.

The closest temperature differential figure in Fig. 14-1 is 15°F; this is the column we will use. Reading down this column, we find that the first pool area (355 square feet) is not quite large enough, so we move to the next column which reads 711 square feet. Reading to the left, we find that we'll need a 160,000-Btu/hr-input heater for either natural or L.P. gas. The *Model No.* column in this particular chart is for swimming pool heaters manufactured by Rudd Manufacturing Co. Any heater providing the same amount of Btu's would be quite satisfactory.

INSTALLING THE HEATER

Once we determine the size of pool heater required for our application, we obtain literature on various models available that will produce the required amount of hot water; check your telephone directory under "Plumbing Supplies" or "Plumbing Contractors." The manufacturer's reputation and the cost of the heater are two items to consider when deciding on a heater.

When the unit arrives, check it immediately for possible damage during shipment and report any claims for damage or shortage immediately.

The unit will look similar to the one shown in Fig. 14-2. Install it in a clean, dry location where a good vent connection can be made. If the unit must be installed outside, it should be

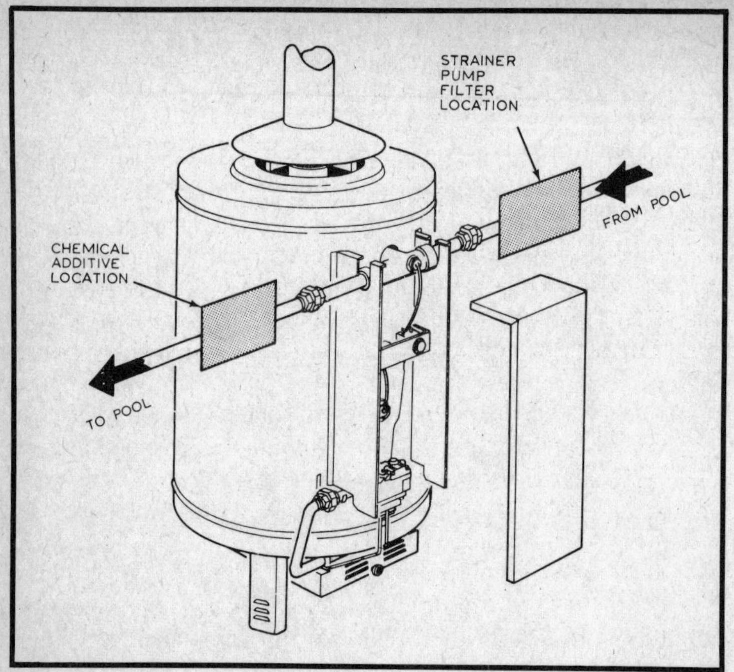

Fig. 14-2. A typical swimming-pool water heater and related connections.

in a sheltered location. A location should be chosen, or provision for protection made, so that if a leak should ever develop, there would be no water damage.

The minimum distance to walls or ceilings to provide adequate clearance for protection of combustible material is six inches from the jacket and draft diverter or vent connector in any direction. However, additional clearance for accessibility to permit servicing such as removing the burner or checking the controls is usually provided. Most models may be used on combustible flooring.

Air for combustion and ventilation must be provided on all gas-fired models. The air supply usually must be through two permanent openings of equal area, one located approximately six inches below the ceiling and one approximately six inches above the floor. Total free area of each opening shall be equal to one square inch per 1000 Btuh of input rating of all appliances in the enclosure.

Where appliances are installed in a confined space within a building of unusually tight construction, air for combustion must be obtained from outdoors or from spaces or ducts freely communicating with the outdoors. Under these conditions, two openings of approximately equal area (one located near the top and one located near the bottom of the enclosure) must be provided with a total free area of not less than one square inch per 4000 Btuh of total input rating of all appliances in the enclosure. If horizontal ducts are used, each opening shall have a free area of not less than one square inch per 2000 Btuh of total input rating of all appliances in the enclosure.

Should the heater be installed near a corrosive, or potentially corrosive, air supply, the heater should be isolated from it and outside air supplied as recommended above.

Examples where potentially corrosive atmospheres could be found are beauty shops and dry-cleaning establishments. The air near such establishments may be safe to breathe, but when it passes through a gas flame, corrosive elements are liberated which will shorten the life of any gas-burning appliance.

Most units come from the factory with controls for operating at a maximum inlet gas pressure of 1/2 psi. If higher pressures are encountered, consult your local gas company for corrective devices.

The gas line should further be of adequate size to prevent undue pressure drop and never smaller than the size of the connection on the heater. Refer to Fig. 14-3 for proper pipe sizing.

A union fitting and a manual shutoff valve should be provided in the gas line near the heater so that the control assembly may be removed easily. The valve should be readily accessible for turning on or off. Also install a trap in the supply line as close to the heater as possible. The pipe itself should be of a material that is resistant to the action of liquified petroleum gases.

When the piping to the unit is completed, check all fittings, valves, couplings, etc., for gas leaks, but do not use an open flame for this; use a soap and water solution.

You will also have to provide a vent of adequate draft capacity and in good usable condition. Most units come with a factory-supplied draft diverter in place which is designed to break chimney drafts that might affect combustion or blow out the pilot light. The venting connectors must be attached to the draft diverter outlet to connect the heater with the gas vent or chimney. If any horizontal run of this connector occurs, it should have an upward slope of at least 1/4 inch per foot.

Maximum Capacity of Pipe in Cubic Feet of Gas per Hour for Gas Pressures of 0.5 Psig or Less and a Pressure Drop of 0.3 Inch Water Column
Based on a 0.60 Specific Gravity Natural Gas; if 2.0 L.P. Gas is used multiply capacity by 0.55

Nominal Iron Pipe Size, Inches	Internal Diameter, Inches	Length of Pipe, Feet													
		10	20	30	40	50	60	70	80	90	100	125	150	175	200
1/2	.622	132	92	73	63	56	50	46	43	40	38	34	31	28	26
3/4	.824	278	190	152	130	115	105	96	90	84	79	72	64	59	55
1	1.049	520	350	285	245	215	195	180	170	160	150	130	120	110	100
1-1/4	1.380	1,050	730	590	500	440	400	370	350	320	305	275	250	225	210
1-1/2	1.610	1,600	1,100	890	760	670	610	560	530	490	460	410	380	350	320

Fig. 14-3. Table giving maximum capacity of pipe in cubic feet of gas per hour.

Where two or more water heaters connect to a common vent (or chimney), the area of the common passageway should be at least equal to the area of the largest vent connector plus 50 % of the areas of any additional connectors. However, there is a definite Btuh capacity for any given vent style and height which should not be exceeded.

An opening is usually provided near the top of all tanks for the installation of a temperature and pressure relief valve. The relief valve is extra protection against damage that could be caused by malfunctioning controls or excess water pressure. If a relief valve is not used, the heater could be extremely dangerous.

Connect the outlet of the relief valve to a suitable drain and use a pipe no smaller than that of the outlet valve. The end of the discharge line should not be threaded (according to one manufacturer) and should be protected from freezing. No valve of any type shall be installed between the relief valve and tank or in the drain line.

The water connections are made similar to those shown in Fig. 14-2. This particular water heater is equipped with a special bypass tee which will probably maintain a correct flow of water through the water heater. Care must be taken to connect the piping from the pump to the connection marked "in" on the bypass tee. Heated water to the pool is then piped from the "out" connection. In this case, both are designed for use with 1 1/2-inch pipe; both, however, may be reduced if desired.

The bypass tee may be rotated 180° to reverse the inlet and outlet connections. In order to allow the thermostat bulb to slip while the tee is being turned, loosen the small nut on the capillary seal fitting.

Follow the manufacturer's recommendations for the sizing of the pumping equipment as the equipment must supply a minimum amount according to the size of heater used. The table in Fig. 14-4 gives the recommended water flow rate (minimum and maximum) for various sizes of heaters. If the flow exceeds the recommended maximum, install a bypass pipe to divert some of the water around the heater so the flow through the heater will be in the recommended range.

OPERATING PROCEDURES

Before lighting the pilot, make sure that the pump has run long enough to fill the tank with water. You can check this by lifting the handle on the relief valve, but don't let the sudden thrust of air or water frighten you.

Once you are certain that the tank is completely filled (allowing some room for air), you may light the pilot light by following instructions provided with each unit. Most units have these instructions printed on the rating plate attached to the front of the heater.

A non-adjustable safety control is mounted in most tanks to limit the tank temperature to 140°F (Fig. 14-5). An adjustable thermostat providing temperatures from 60° to 95°F should be installed on the unit if it does not have one when it comes from the factory. This thermostat may be set to produce any desired temperature in this range; 78°F is the temperature most often chosen.

Most heaters are designed to operate at a pressure equivalent to 3 1/2 inches of water (for natural gas) or 11 inches of water (for L.P. gas). Most heaters have a 1/8-inch

HEATER INPUT BTUH	RECOMMENDED FLOW RATE GPM	
	MINIMUM	MAXIMUM
108,000	13	38
132,000	15	46
156,000	18	55
199,900	23	70

Fig. 14-4. Chart of recommended flow rate (GPM) for various sizes of heater input (Btuh).

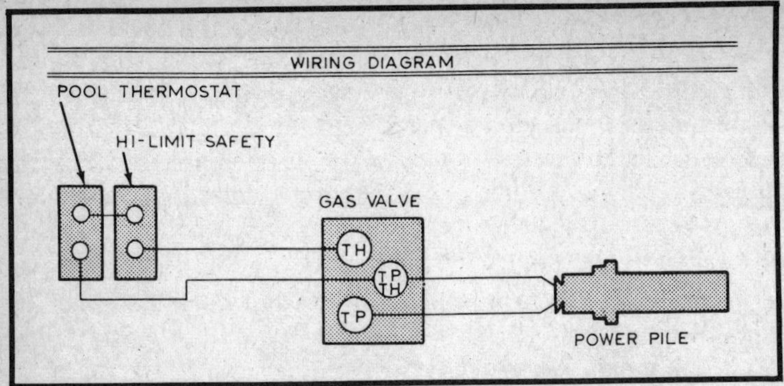

Fig. 14-5. Wiring diagram of typical pool-heater control.

NPT tapping provided on the manifold for connecting a gauge to check this pressure. Most natural gas models are furnished with a preset gas pressure regulator while L.P. models use the service regulator on the gas company's tank.

You should also check the input of the heating unit against that which is shown in the nameplate rating. You can consult your gas company to determine the heating value of the gas supplied; then check the input by clocking the gas meter while all other gas appliances on the meter are turned off (including their pilots). Use the following formula.

$$\text{Input} = \frac{3{,}600 \times \text{heating value} \times \text{number of cubic feet timed}}{\text{seconds clocked}}$$

To insure accuracy of the rating, the clocked time should be at least 60 seconds.

Minor adjustments in input can be made by varying the manifold pressure from the designated settings. Any large change should be made by changing the burner orifice size with the advice of your local gas company. You should also consider that the ratings of the heaters are based on sea level operation and are fairly accurate for installations at elevations up to 2000 feet. For elevations above 2000 feet, the rating is reduced 4% for each 1000 feet above sea level.

To adjust the pilot flame, remove the capscrew cover and adjust the gas flow to the point where the thermopile tip is completely enveloped by the flame; then replace the cap. Some models have a pilot pressure regulator and require no adjusting.

MAINTENANCE

An access hole is provided on all swimming pool water tanks for the purpose of cleaning out the tank. Good maintenance practice requires that the tank be frequently drained, inspected, and cleaned of deposits. Foreign material collects and, unless the water supply is naturally soft, scale or lime deposits will accumulate in the tank. Accumulation of these deposits can shorten the life of the water heater and, in excess, can void the warranty. To clean or inspect one type of tank, proceed as follows:

1. Shut off the gas valve and drain the tank.
2. Remove the jacket cover panel and, with a pocket knife, cut and remove a circular plug of insulation the full size of the jacket opening.
3. Remove the flue liner seal plate.
4. Loosen the nut on the seal plate assembly and twist the yoke sideways.
5. Hold the assembly securely and push it inward to loosen, then twist it sideways and remove it from the tank.
6. A new gasket should be installed on the seal plate before it is reinstalled on the tank.

When cleaning, care should be taken that the tool used does not damage the interior surface of the tank. Remove any deposits 1/16-inch or greater in thickness.

Periodic cleaning of the equipment is also advised. Remove all dust from under the heater and brush off the underside of the base. Remove any tank scale that may have fallen on top of the burners. Also remove any lint or dust that may block the free passage of air to the burner and combustion chamber. It is advisable to have the heater thoroughly cleaned and the controls checked once a year by a qualified repairman.

During the winter, when the pool is not in use, turn the shutoff valve to the "OFF" position and drain all water from the tank. Be sure to fill the tank with water in the spring before relighting.

Chapter 15

Ole Sol

The use of solar energy is really nothing new. Man has depended upon it for thousands of years to provide light for work and play; heat to dry and preserve foods, heat for cooking or to warm dwellings; and to help build homes (as in the case of sun-dried adobe bricks used in dwellings in the southwestern section of North America). Now our energy shortage is forcing man to harness our greatest source of energy for other useful purposes.

Solar energy is readily available, well-distributed, inexhaustible for all practical purposes, and does not pollute the environment. The sun has been used to power space heaters, pumps, sewage treatment plants, steam engines, and household appliances. In fact, we can now produce all the forms of energy necessary for daily living by using solar energy.

One the most promising applications of solar energy is heating homes. Dr. Harry E. Thomason was one of the early experimenters in the use of solar heat for residential heating. An article in the February 1965 issue of *Popular Mechanics* gave the operating principles of a home he designed which was heated and cooled by a then-unique system combining solar energy and water. Since that time, Dr. Thomason and his family have built three solar-heated homes that have been consumer-proven for over 10 years. Dr. Thomason's designs are available from Edmund Scientific Co., 701 Edscorp Building, Barrington, N.J. 08007.

His designs are sound and well-proven; you may use them for guidance or as inspiration for yourself or your architect. However, a license from Dr. Thomason is necessary if you want to utilize his exact drawings. At this writing, they cost $40 and are also available from Edmund Scientific Co.

Another system, designed by G. Lewis Craig & Associates of Waynesboro, Va., is a very practical design for the use of solar energy. A description of this system follows:

The building for which this solar heating system was designed is a relatively small (1125-square-foot) structure with a full-size basement. The above-grade structure consisted of wood framing with brick veneer. Figure 15-1 shows the basement layout with a 2000-gallon storage tank, the air-handling units, the ductwork for the system, and the auxiliary heating systems; that is, heat pump, electric duct heater, etc. The isometric piping diagram is shown in Fig. 15-2.

To describe normal operation, the 2000-gallon tank is first filled with water. Thermal sensors located on the face of the solar collector plate and in the water storage tank sense the temperature of the water. If the temperature is below the preset rating, the solar collector pump (No.1) is energized and begins pumping water from the tank through the 1 1/4-inch pipe to the uppermost part of the solar collector located on the roof of the building (as shown in Fig. 15-3). The water then runs—by gravity—through the collector plates to a 1 1/4-inch pipe which carries the heated water back to the water storage tank.

If the temperature of the water has reached its preset rating, and the building thermostat calls for heat, pump No. 2 begins operating, and circulates the heated water through the heat exchanger which—in this case—is a heat pump. This heat pump takes the heat from the water and distributes it through the structure by means of a forced-air system. However, since heat is removed from the water, it becomes cool and therefore must be recirculated through the solar collector for reheating before it circulates through the heat pump again.

During the summer months, the heat-pump cycle is reversed for cooling the building. The hot-water tank is not used during this cycle; instead the water is diverted to an outside cooling tower to cool it. Pump No. 3 circulates the water during this operation, and various automatic valves control the direction of water flow. The location of the water cooler (cooling tower) is shown in Fig. 15-1 and two sections of

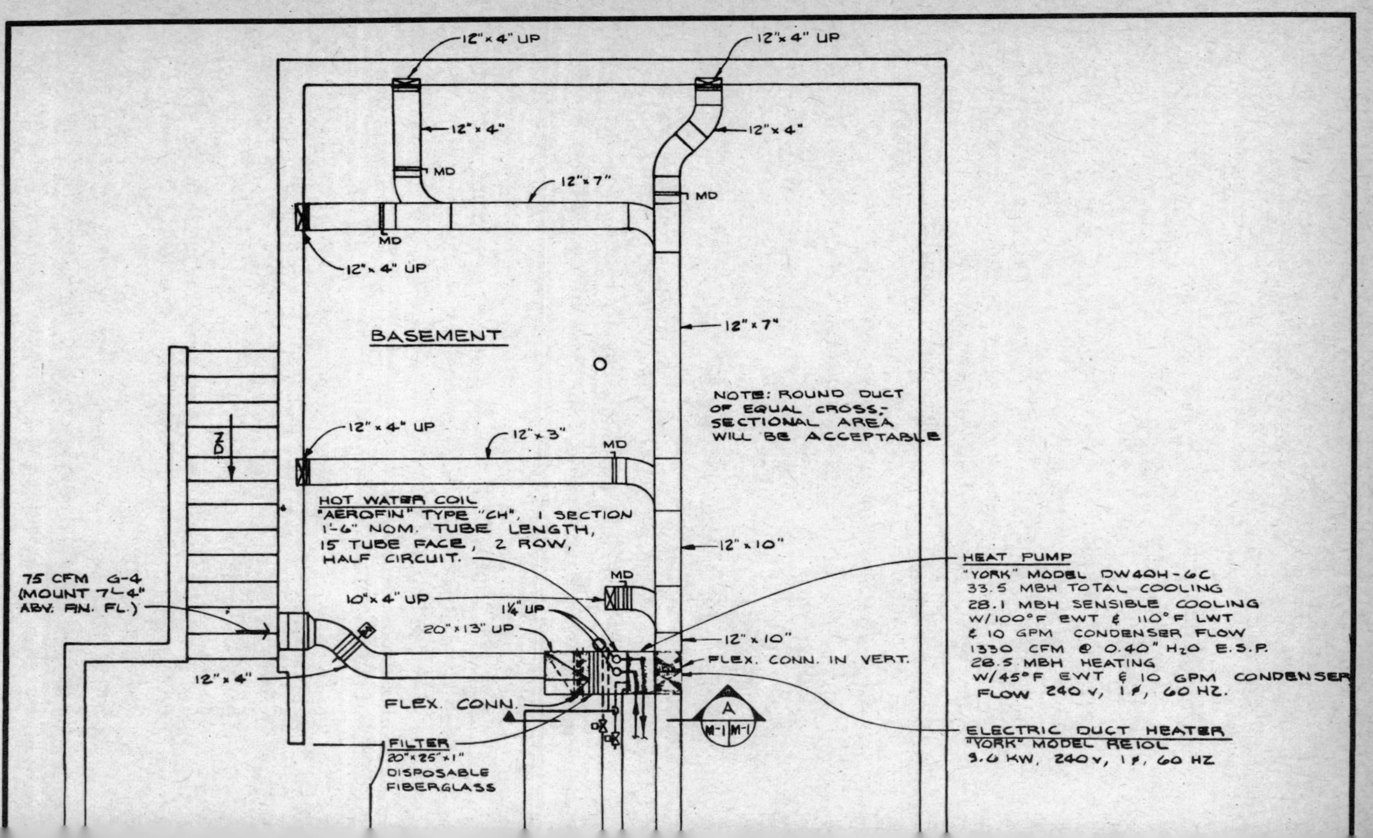

12"x4" UP
12"x4" UP
12"x4"
12"x4"
MD
12"x7"
MD
MD
12"x4" UP
BASEMENT
12"x7"
NOTE: ROUND DUCT
OF EQUAL CROSS-
SECTIONAL AREA
WILL BE ACCEPTABLE
12"x4" UP
12"x3"
MD
HOT WATER COIL
"AEROFIN" TYPE "CH", 1 SECTION
1'-6" NOM. TUBE LENGTH,
15 TUBE FACE, 2 ROW,
HALF CIRCUIT.
12"x10"
MD
75 CFM G-4
(MOUNT 7'-4"
ABV. FIN. FL.)
10"x4" UP
1¼" UP
20"x13" UP
12"x10"
FLEX. CONN. IN VERT.
12"x4"
FLEX. CONN.
FILTER
20"x25"x1"
DISPOSABLE
FIBERGLASS
A
M-1 M-1
HEAT PUMP
"YORK" MODEL DW40H-6C
33.5 MBH TOTAL COOLING
28.1 MBH SENSIBLE COOLING
W/100°F EWT & 110°F LWT
& 10 GPM CONDENSER FLOW
1330 CFM @ 0.40" H2O E.S.P.
28.5 MBH HEATING
W/45°F EWT & 10 GPM CONDENSER
FLOW 240 V, 1 ø, 60 HZ.
ELECTRIC DUCT HEATER
"YORK" MODEL RE10L
3.6 KW, 240 V, 1 ø, 60 HZ

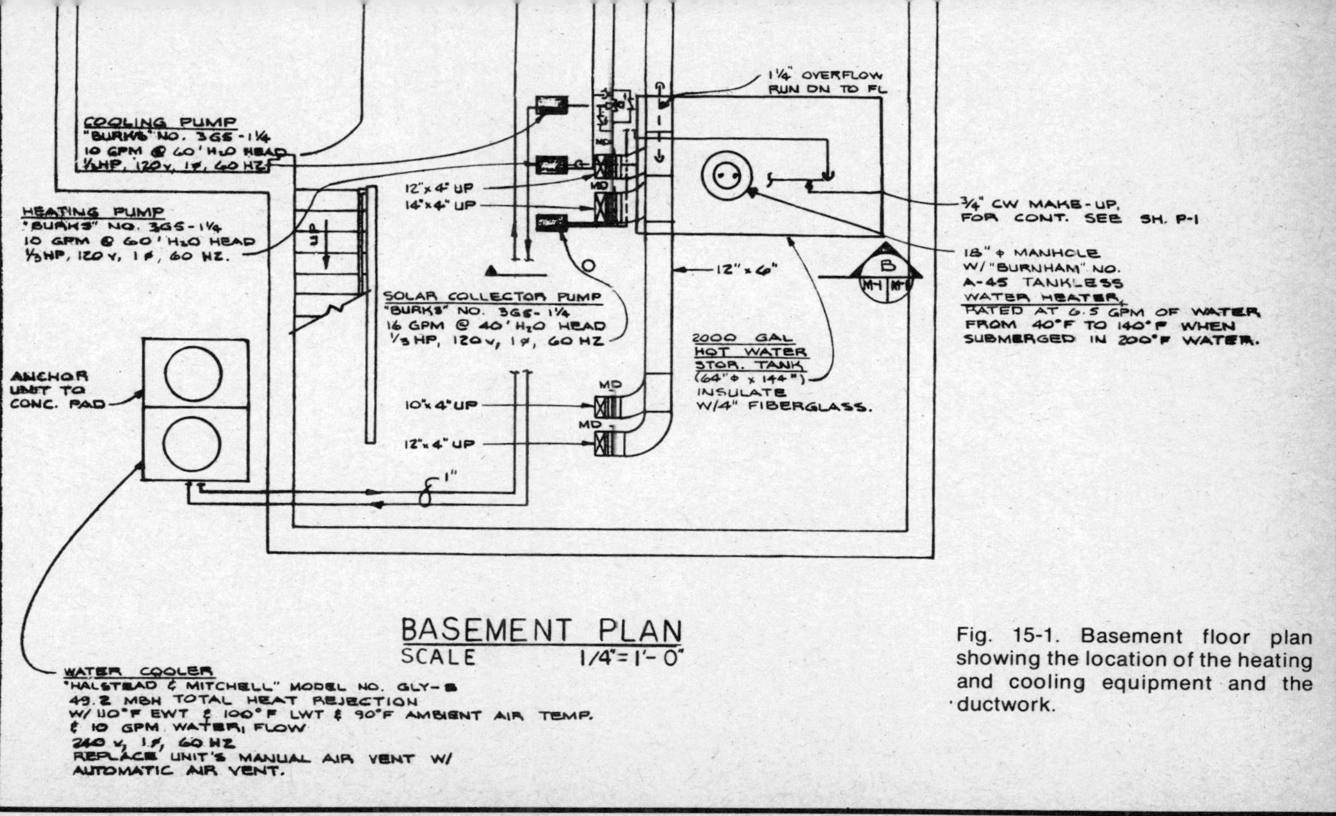

Fig. 15-1. Basement floor plan showing the location of the heating and cooling equipment and the ductwork.

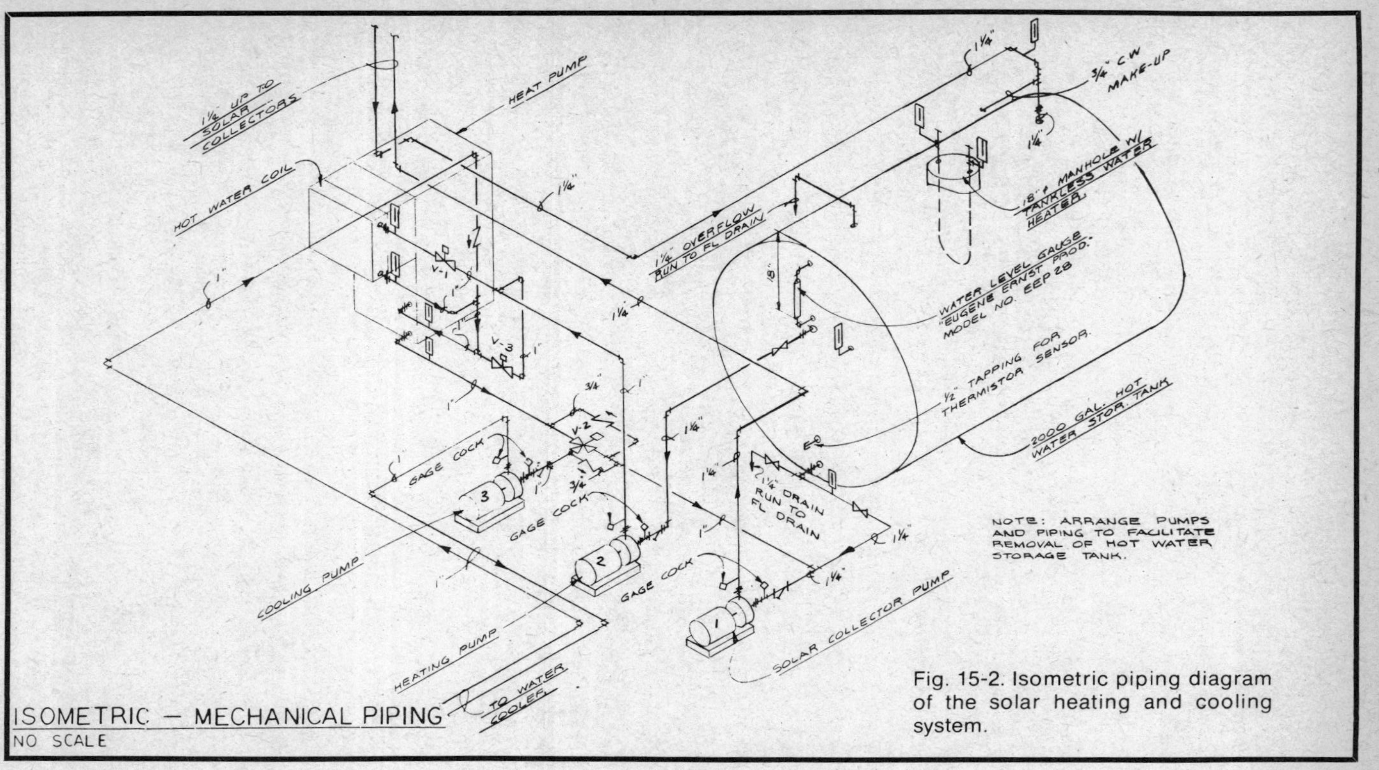

Fig. 15-2. Isometric piping diagram of the solar heating and cooling system.

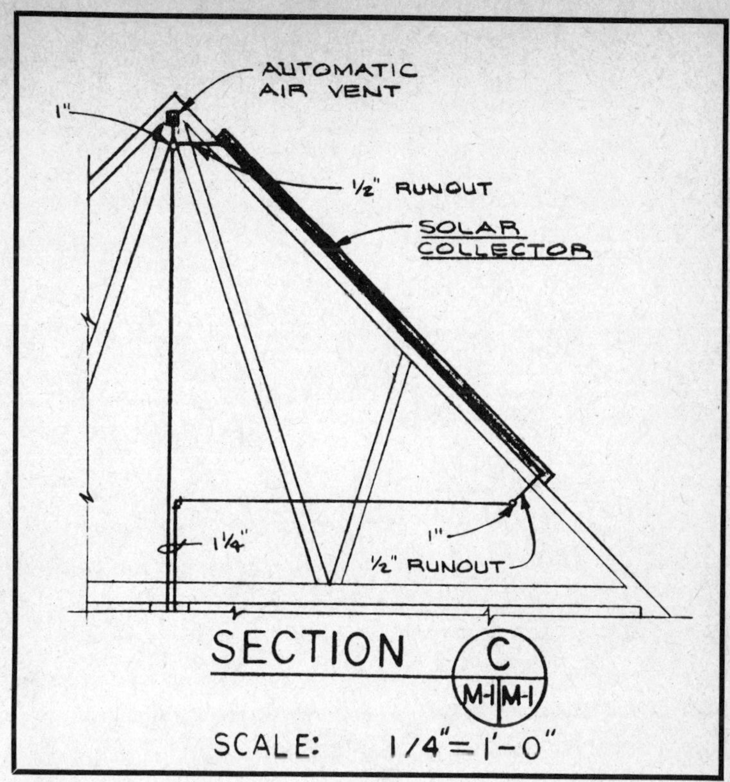

Fig. 15-3. Section through roof showing the location of the solar collector, piping, and automatic air vent.

the heat pump, related ductwork, and the water storage tank are shown in Figs. 15-4 and 15-5. Control wiring for the heat pump, solar collector, and outside air damper are shown in Figs. 15-6, 15-7, and 15-8, respectively.

A section through the roof showing the location of the solar collector panel and other details of construction is shown in Fig. 15-9. Notice the catwalk along the bottom edge of the solar collector: a walkway to facilitate collector maintenance and repair. A detail of this catwalk is shown in Fig. 15-10. Notice that the pressure-treated 2 × 10 inch plank is secured to a 2-inch flange by means of 1/2-inch bolts. This flange is in turn secured to a 1/4-inch steel support on 4-foot centers, and bolted to the 1/2-inch sheathing with 1/2-inch bolts.

A plan view of the collector is shown in Fig. 15-11 while the various sections (A, B, and C) are shown in Figs. 15-12, 15-13, and 15-14.

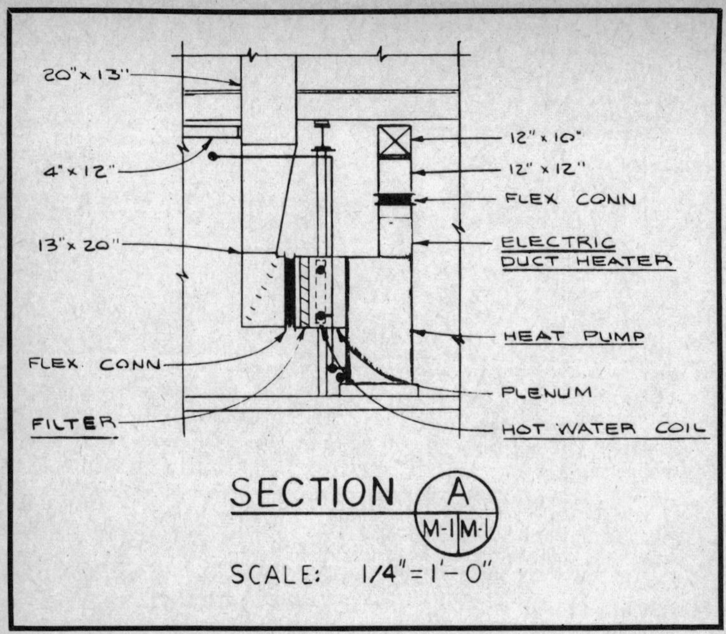

Fig. 15-4. Section "A" of the floor plan in Fig. 15-1.

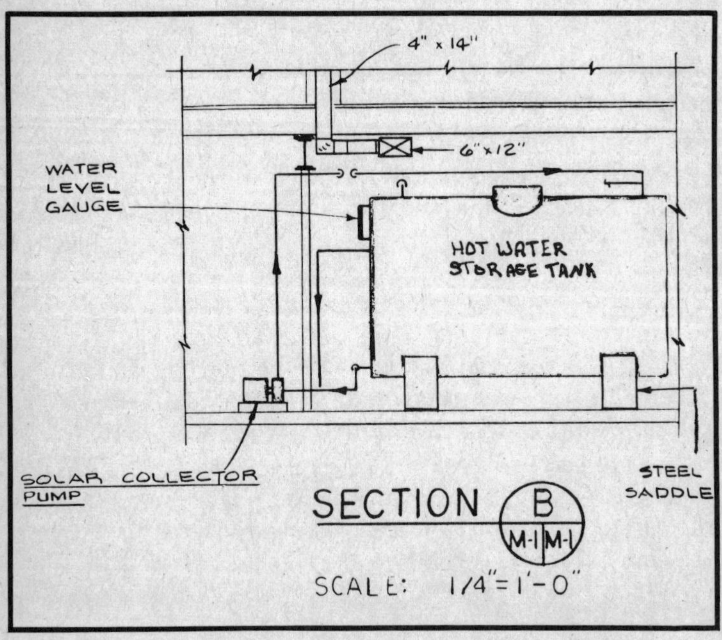

Fig. 15-5. Section "B" of the floor plan in Fig. 15-1.

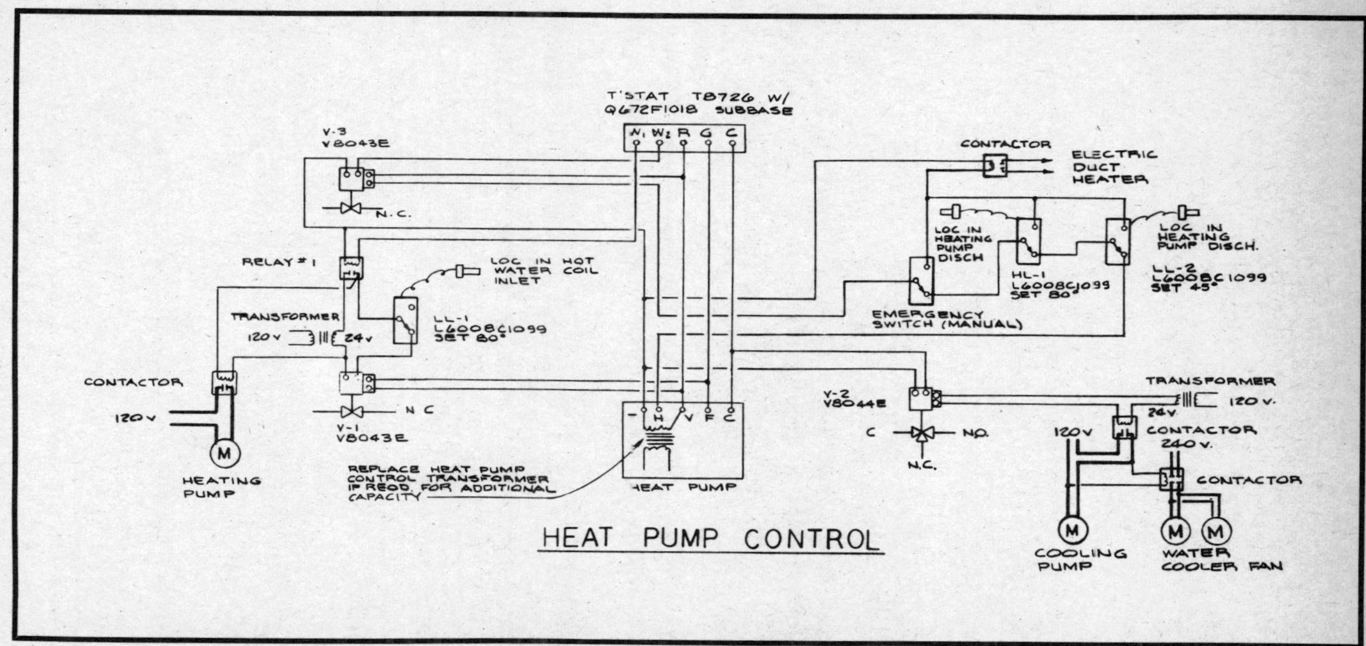

Fig. 15-6. Control wiring diagram for the heat pump described in the text.

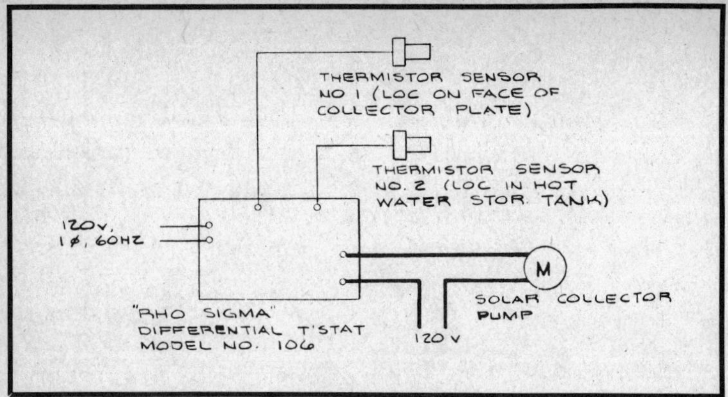

Fig. 15-7. Solar collector control diagram.

The first winter's use of this system proved to be entirely satisfactory and a substantial savings in fuel was realized, although it has not been determined just how long it will take for the fuel savings to offset the higher installation cost of this system. The initial cost of a solar system is at least twice that of a conventional hot-water or forced-air heating and cooling system.

There are ways in which the homeowner can cut this initial expense, however: one way is to do most of the work himself after the system has been carefully designed by a qualified architect or engineer.

In the project just discussed, the water cooler could have been eliminated, with its function performed by having the water run over the opposite side of the roof during the night.

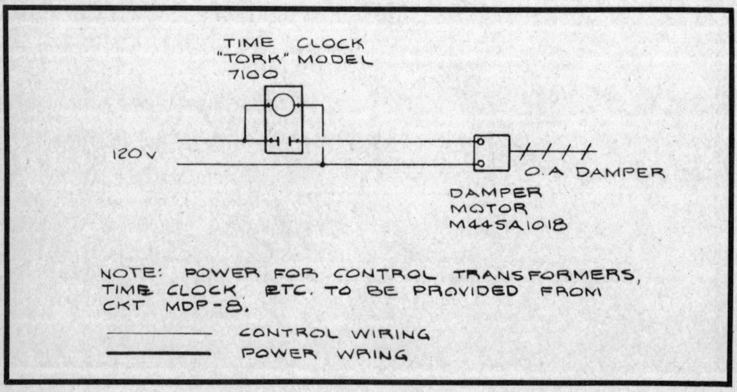

Fig. 15-8. Outside air damper control diagram.

While not as efficient as the water cooler, this method would decrease the initial cost considerably as well as summer operating costs.

Circulating the water from the tank over the opposite roof at night would chill it by the processes of radiation and evaporation. During the day, the heat pump—on its cooling cycle—would then transfer the home's heat to the cool water. To ensure that the water in the system didn't get too warm to absorb heat efficiently, you could install a small refrigerating unit in series with the water piping on the discharge side of the tank.

As you were looking over the drawings and other details of this solar system, you may have noticed the heater in the ductwork. Why an electric heating element in a system supposedly heated by the sun? The is to assure the owners of sufficient heat during successive cold, cloudy days when little solar energy is available. When this occurs, and the heat pump is unable to obtain enough heat from the water to keep the thermostat satisfied, the auxiliary electric duct heaters energize to supplement the heat pump.

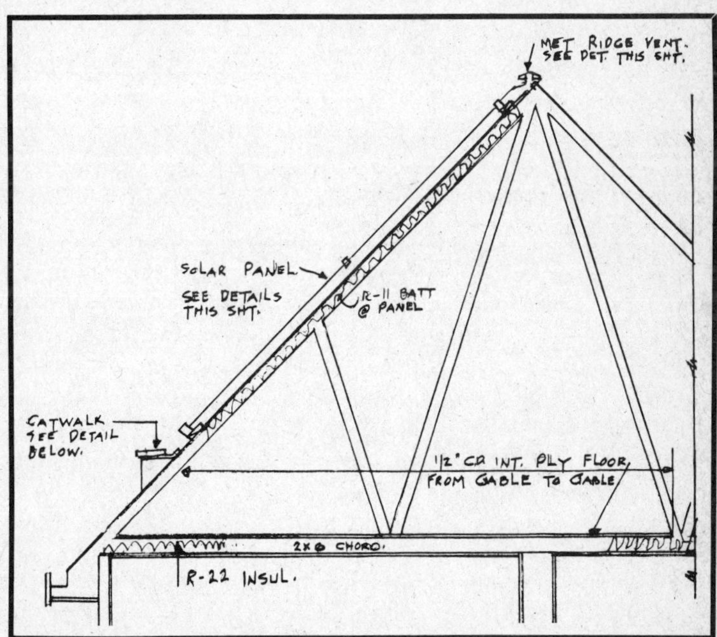

Fig. 15-9. Section through roof showing the location of the solar collector panel, catwalk, and other details of construction.

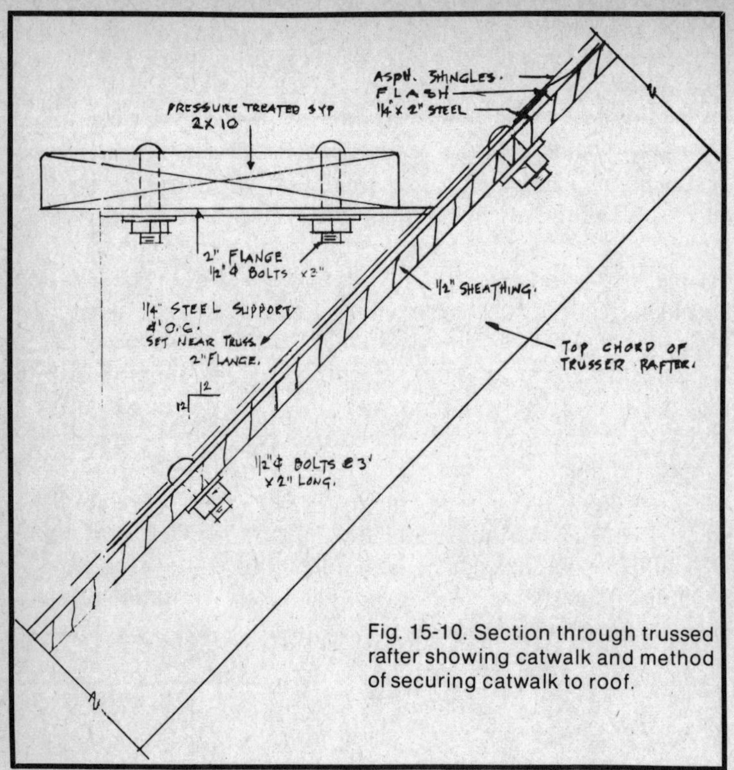

Fig. 15-10. Section through trussed rafter showing catwalk and method of securing catwalk to roof.

This is only one example. There are many different money-saving, pollution-free ways to heat your home, swimming pool, driveway, and walks with solar energy.

We are sure to see more and more solar-heated homes in the future, so why don't you get started right now on yours. Pick up a set of plans from Edmund Scientific Co. along with a few of the books explaining in detail how several different types of solar heating systems operate. Study these details carefully, and then incorporate those that suit your fancy in a design of your own. Then you might want to seek the assistance of a qualified engineer to iron out the rough spots. With complete working drawings and specifications, you should have little trouble doing most of the work youself with only conventional hand tools.

Perhaps you don't want to heat your entire home with solar energy, but you would like to utilize some type of solar system to reduce your utility bills. One possibility is a solar-heated domestic hot-water system.

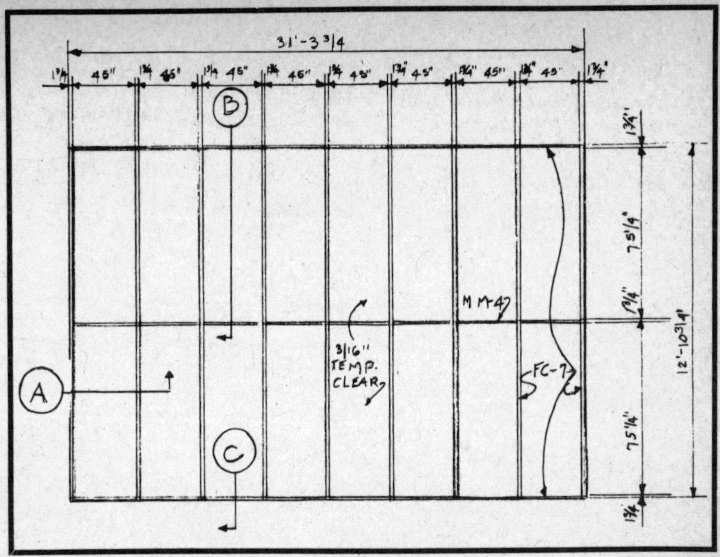

Fig. 15-11. Collector glazing plan.

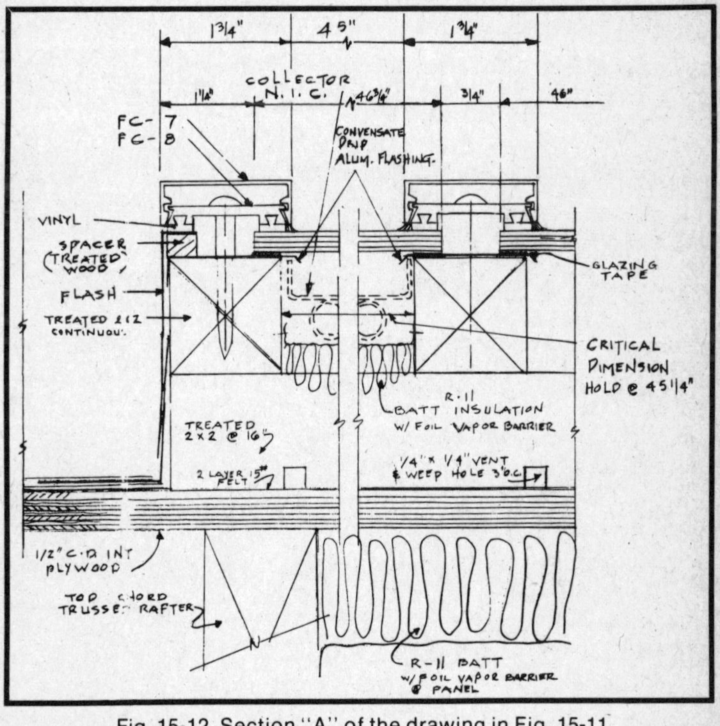

Fig. 15-12. Section "A" of the drawing in Fig. 15-11.

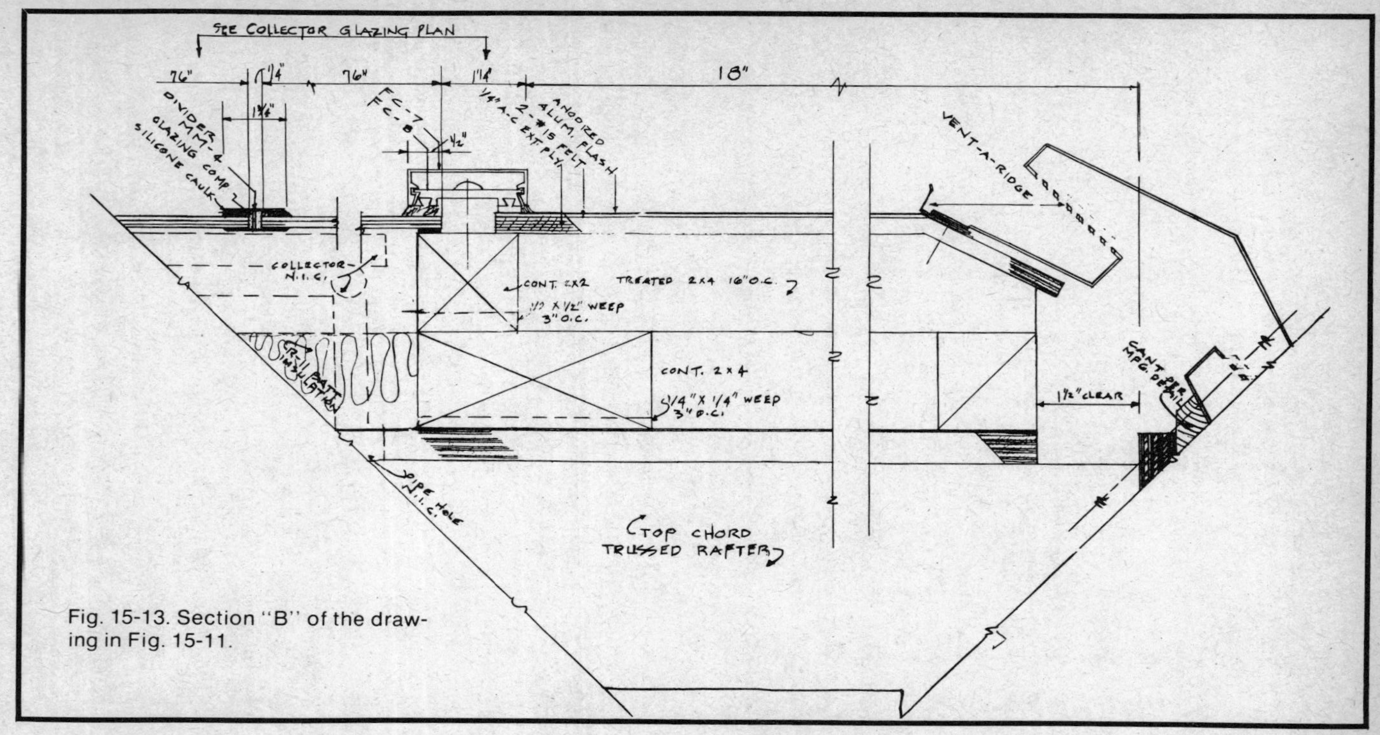

Fig. 15-13. Section "B" of the drawing in Fig. 15-11.

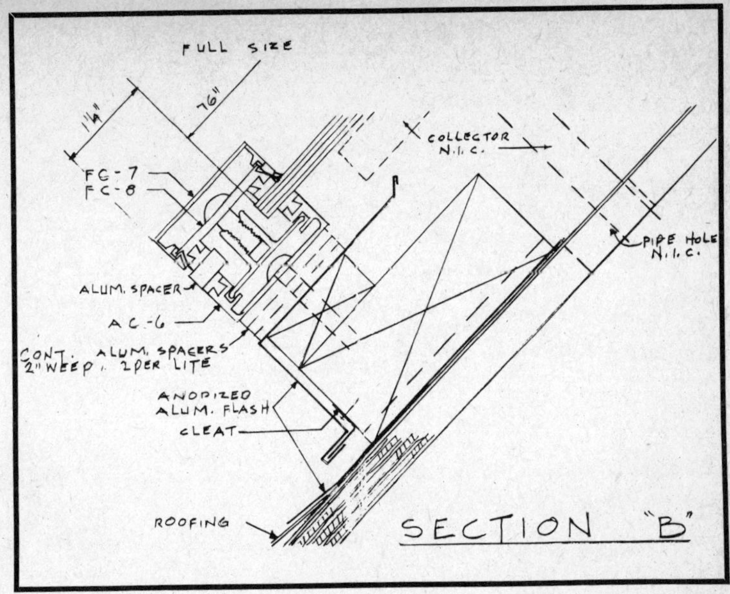

Fig. 15-14. Section "C" of the drawing in Fig. 15-11.

Again, there are many ways to go about this. One way is to install a pressurized water tank inside a larger tank, allowing the solar-heated water to circulate around the smaller tank, heating the domestic water inside it. The solar collector itself can consist of black corrugated aluminum roofing backed with insulation and covered with clear plastic or glass panels. A pump will be required to pump the water to the ridge of the collector. From the ridge it trickles down the collector which heats it. Experiments with this type of system have proved that water can be brought to the boiling point in two or three cycles across the collector.

A storage tank for the heated water provides the best results, but even just circulating the water over the solar collector and then around the hot water tank will produce substantial savings on your water heating bill. However, if a large storage tank of—say, at least 1000 gallons—is not used, it is advisable to connect a conventional electric or gas water heater in parallel with the hot water tank around which the solar-heated water flows. Then if the solar system cannot handle the load on cloudy days, the electric elements on the auxiliary heater will energize and provide the additional heat required.

Chapter 16

Basic Safety Controls for Hot-Water Space-Heating Boilers

Every hot water boiler must have a safety device so that it may be operated without possible damage to the boiler or to persons or property near the boiler. A safety relief valve that will keep the pressure at or below the maximum allowable working pressure should be installed on every boiler but, until recently, the methods of accomplishing this were not clearly understood.

Today the A.S.M.E. Boiler Code defines the correct procedure in detail. Despite this clear-cut explanation, some installations are still being made in the old-fashioned haphazard way. So if your home is heated with a hot water boiler or you are planning to have one installed, the information contained in this chapter will help you to make certain that the installation is a safe one.

Figure 16-1 shows an *unsafe* method of attempting to provide protection against overpressure:

1. The relief valve does not comply with the A.S.M.E. Boiler Code requirement.
2. Its capacity is unknown.
3. It is installed in the wrong location.
4. It can inadvertently be isolated from the boiler by lime or scale build-up in the boiler feed line.
5. The function of a relief valve has nothing in common with a pressure-reducing fill valve. A combination of the two units is based on price considerations—not performance.

The first basic step in providing correct safety control for hot water boilers is to make sure that the relief valve complies with the A.S.M.E. Boiler Code. Following the Code provisions assures adequate protection against overpressure developing. The installation in Fig. 16-2., for example, shows the same boiler as in Fig. 16-1—this time with a safe installation.

The valve is an approved relief valve, constructed in accordance with A.S.M.E. Boiler Code requirements; it has a rated capacity certified by the National Board of Boiler and Pressure Vessel Inspectors. A try lever on the valve provides a means of testing it and the relief valve and its related piping are installed on top of the boiler, in accordance with Code procedures.

So remember the first basic step: safety demands a certified, Btu-rated relief valve, correctly installed and matched to the gross output of your boiler. If you don't know

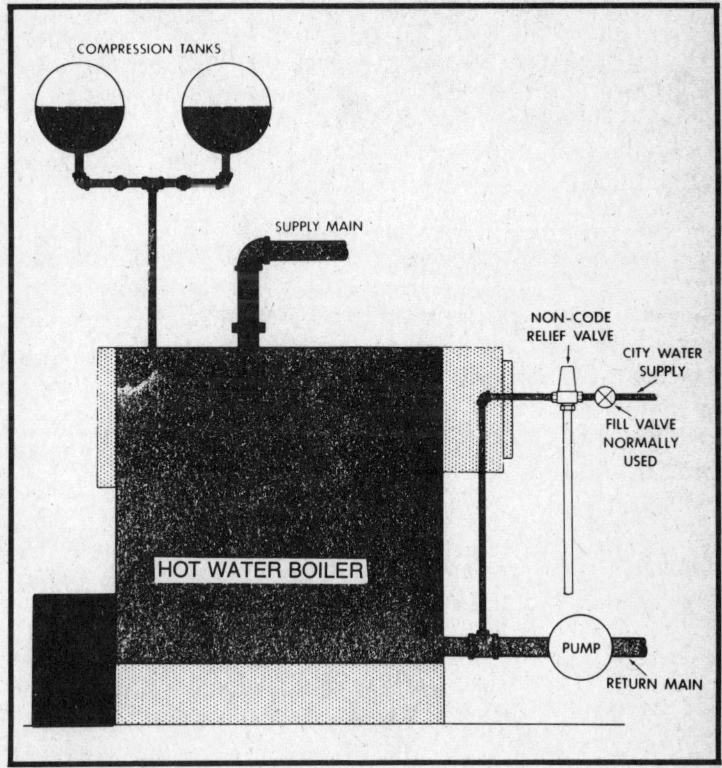

Fig. 16-1. Boiler piping diagram showing unsafe method of attempting to provide protection against overpressure.

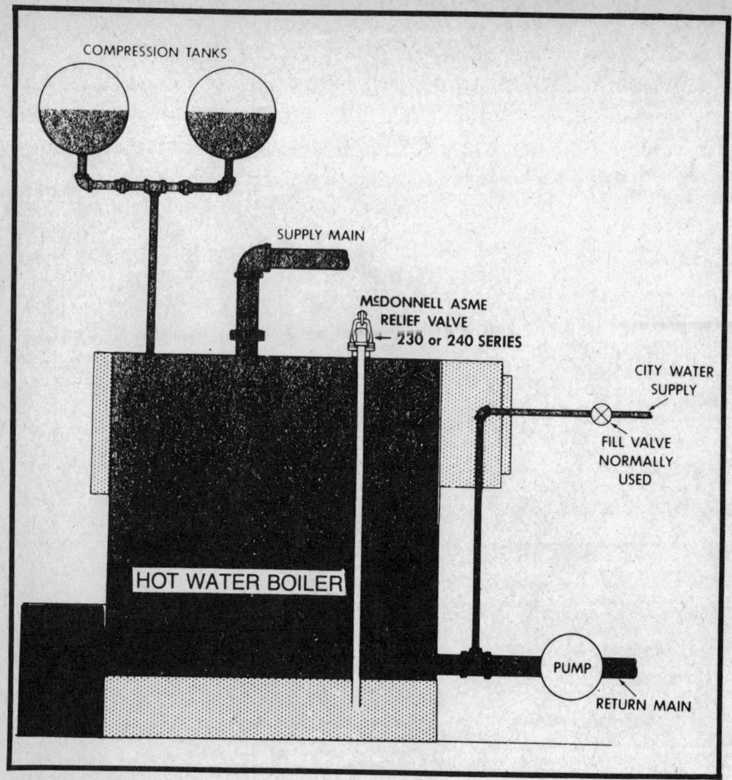

Fig. 16-2. Same boiler as shown in Fig. 16-1 with a correct installation.

its gross output, look for a nameplate on your boiler which will give all of this information.

The average hot-water boiler is constructed for a maximum working pressure of 30 psi; its safety relief valve should be set to open at this pressure. Therefore any circumstance which will elevate the boiler pressure to 30 psi will cause the relief valve to open.

If your boiler safety relief valve opens frequently, you probably have a problem in your system. Here are a few of the reasons a safety relief valve opens:

1. Hand-filling the boiler and system and allowing the full city water supply pressure to act against a full system. The city water pressure could be from 40 to 60 psi or even higher.
2. Hydrostatic testing of a system with pressures in excess of 30 psi.

3. Waterlogged compression tanks which eliminate the space required for thermal expansion of water.
4. Undersized compression tanks not adequate for thermal expansion.
5. Excessive static head or pump discharge pressures.

In all of the instances listed above, the relief valve discharges water since the boiler water temperature is normally below 212°F. However, in some emergency cases—like over-firing the burner—a more critical demand is placed on the safety relief valve: it must discharge both high temperature hot water and steam. Whenever the water temperature in the boiler is above 212°F and the relief valve opens, the sudden pressure drop causes the superheated water to flash to steam. Therefore, the discharge capacity of an A.S.M.E. relief valve is tested and rated on steam.

As mentioned previously, an emergency condition could be caused by an over-firing of the burner; that is, the heating system cannot dissipate the heat energy as fast as it is developed in the boiler, and temperature and pressure continue to rise. If this occurs in your boiler, check for the following:

1. Failure of the limit control to stop the burner.
2. Mechanical failure of a fuel valve.
3. Burner set on manual operation.
4. Residual heat (with coal-fired equipment).
5. Burner considerably oversize in relation to the boiler system.

In the instances listed above, the safety relief valve discharges steam. If make-up water does not replace the water lost through the relief valve, a hazardous low-water condition can result. In fact, the record clearly indicates that most hot water boiler damage can be traced to low water.

The construction of a hot water boiler and a steam boiler is essentially the same. Most of the causes of low water in one will also hold true for the other.

It is a common misconception that a pressure-reducing valve, used to fill a hot-water system initially, will keep the boiler and system full under all circumstances. But when it is realized that a pressure-reducing valve is normally set at 12 pounds or 18 pounds, and a safety relief valve opens at 30 pounds and closes at 26 pounds, it becomes obvious that the

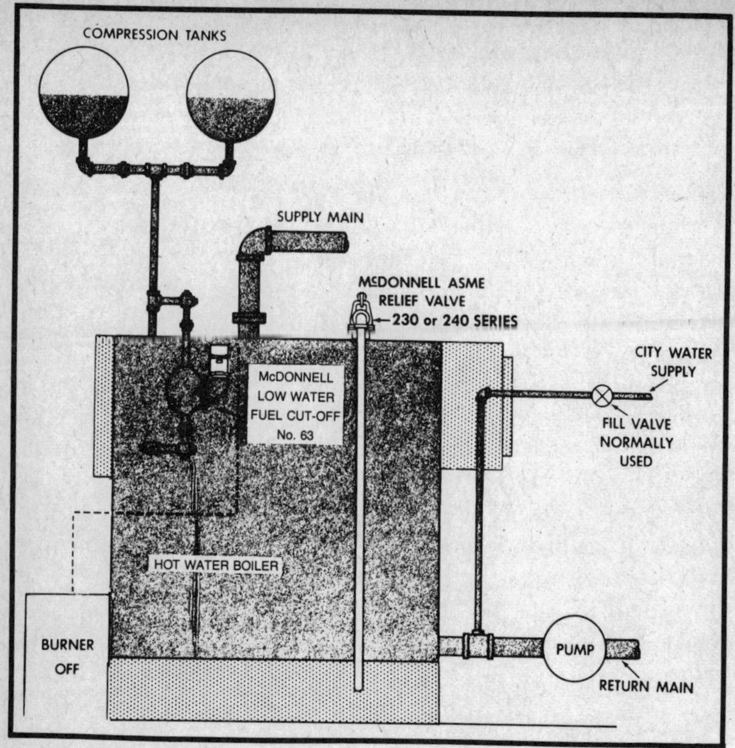

Fig. 16-3. Typical installation of a low water cut-off control on a hot water boiler.

pressure-reducing valve is ineffective during the time the relief valve is functioning.

If a manual fill valve is used, any leak in the system can quickly cause a low-water condition. Sometimes the loss of water is due to carelessness, such as draining the boiler for repair or summer lay-up without eliminating the possibility of firing. When the relief valve opens because of over-firing as mentioned previously, a loss in boiler water can occur.

From this, we can see that a low-water safety control on all hot-water boilers is in order. Here again the similarity between a hot water boiler and a steam boiler is apparent. The minimum basic control that should be installed on a hot water boiler is a float-operated, low-water fuel cutoff. Because of the higher working pressures of a hot water boiler, the low water cutoff is designed for working pressures in excess of 30 psi.

Figure 16-3 shows a typical installation of a low-water cut-off control on a hot-water boiler. There are many possible locations and piping arrangements available for this type of installation. Because there is no normal water line to be maintained in the boiler, any location of the control above the lowest permissible water level is satisfactory.

With a low water safety control added to your boiler, it operates in the usual manner, yet is safeguarded against the emergency condition of low water.

Figure 16-4 shows the action of the low-water cutoff when an emergency condition arises. The lowering of the boiler water line and the simultaneous lowering of the water line in the float chamber causes the float to drop, thus opening the electrical circuit and stopping the automatic burner.

Here, then, is another basic safety control, one that provides a means of stopping the automatic firing device if the water in the boiler drops below the minimum safe level. All boilers should be provided with some type of low-water cutoff.

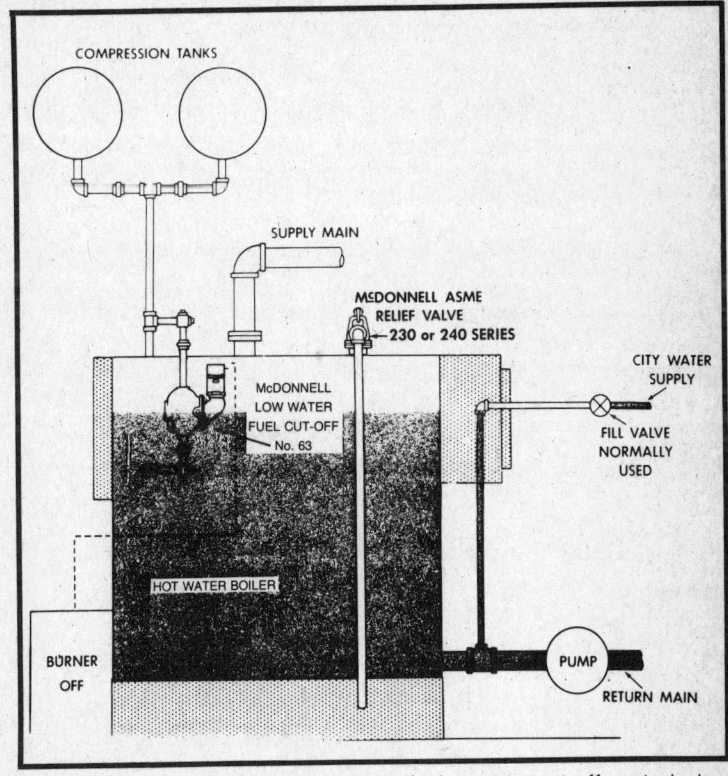

Fig. 16-4. Drawing showing the action of a low-water cut-off control when an emergency arises.

If we could rely absolutely on the low-water cutoff to stop the automatic burner each time a low-water condition developed, then the problem would be solved completely. However, experience has proved that under certain circumstances, the low-water cutoff cannot fulfill its duties.

The safest installation uses an approved combination water feed control and low-water cut-off control. Such a control feeds water to the boiler as fast as it is discharged through the relief valve and, at the same time, stops the electrical operation of the burner when low water occurs. This combination of mechanical and electrical safeguards is the best and most complete recommendation for a residential hot-water boiler.

There are several types of over-firing that can occur which low-water cut-off controls cannot protect against. In such an emergency, the only way the boiler can remain safe is to have water fed into it as fast as it is discharged through the relief valve. Some of these emergency conditions are:

1. Over-firing (by hand) a coal-fired boiler.
2. A fuel feed valve stuck open.
3. The burner system manually set in the ON position.
4. Closing of zone controls, isolating the boiler from the system, with a residual fuel bed in the boiler.

However, should any of the above conditions occur on a boiler with a combination control installed, the operation of the feeder will keep enough water in the boiler to keep it safe.

Therefore, if over-firing exists, the feeder will maintain a minimum safe water level. When the water line in the boiler or system is lowered to the level of the feeder, the float in the feeder drops and opens the feed valve in the make-up water supply line. Should a large leak develop, or should an attempt be made to fire the boiler without adequate water in it, the low-water cut-off switch will prevent the burner from operating.

In many larger heating systems, open expansion tanks are located above the highest point of each circulating zone to keep the system flooded, and to take care of thermal expansion and contraction of the water. Water line regulation in these tanks is essential. It is also desirable to include a high- and low-level alarm to indicate if maximum or minimum tank levels have been exceeded.

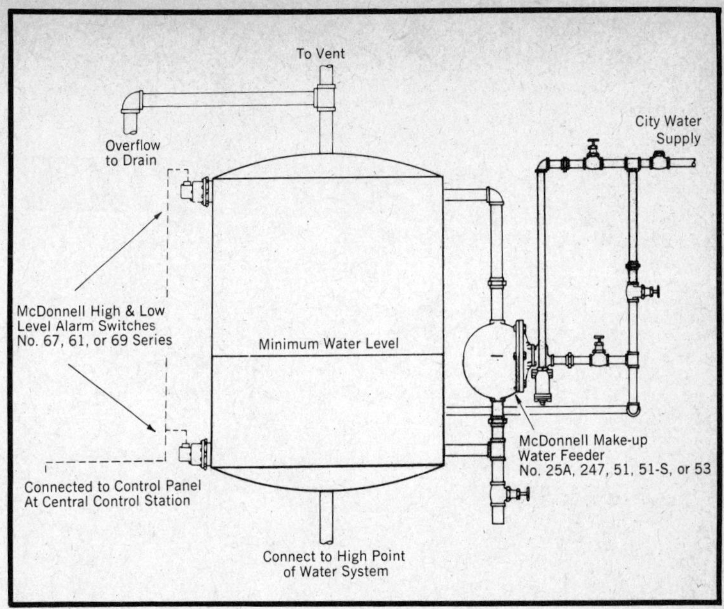

Fig. 16-5. A float-actuated water-feeder control installed on an expansion tank.

The desired minimum water level is maintained by a float-actuated feeder control installed on the expansion tank as shown in Fig. 16-5. Selection of the feeder depends upon the make-up supply pressures and the feeding rate required.

A float-actuated cutoff, installed at the high and low levels of the tank and connected to the central control panel, will signal if either of these limits has been exceeded. On smaller residences, a simple bell or similar sounding device can be used to signal the user when these conditions occur.

In some heating systems, heating units which draw in fresh air from the outside and blow it across a heat-transfer coil through which hot water is circulated are employed. If, during cold weather, circulation of the hot water through the coil should fail for any reason, an unwanted blast of cold air will result and the coil itself may be in danger of freezing.

A float switch can be installed in the hot water line to guard against such a situation and such a switch can be wired to perform the following tasks if flow of hot water stops:

1. Control a motorized damper on the air intake to shut down the flow of cold air to the coil.
2. Stop the fan or blower motor.

3. Sound an alarm or turn on a signal light at a central location in the home.

In some hot-water boiler installations, the return water may be considerably cooler than the boiler water. This may cause rapid changes within the boiler known as thermal shock. Such a situation can occur when a system is started up, after an extended power failure, when zone pumps operate intermittently, or at any other time when circulation through the boiler may be stopped.

Controlling the circulation of water through the boiler is basic to preventing thermal shock. A flow switch installed as illustrated in Fig. 16-6 will measure this water circulation and provide the following means of control:

1. Detect when circulation stops and break the circuit to the firing device until circulation is restored.
2. Detect when circulation stops and operate a three-way valve to increase bypass capacity and reduce circulation through the boiler. When circulation is

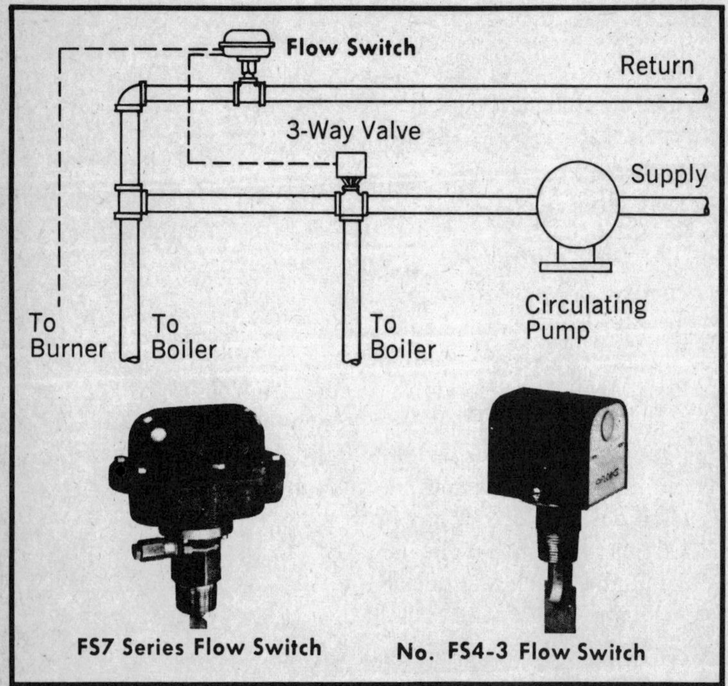

Fig. 16-6. A flow switch used to measure the water circulation and provide a means of control.

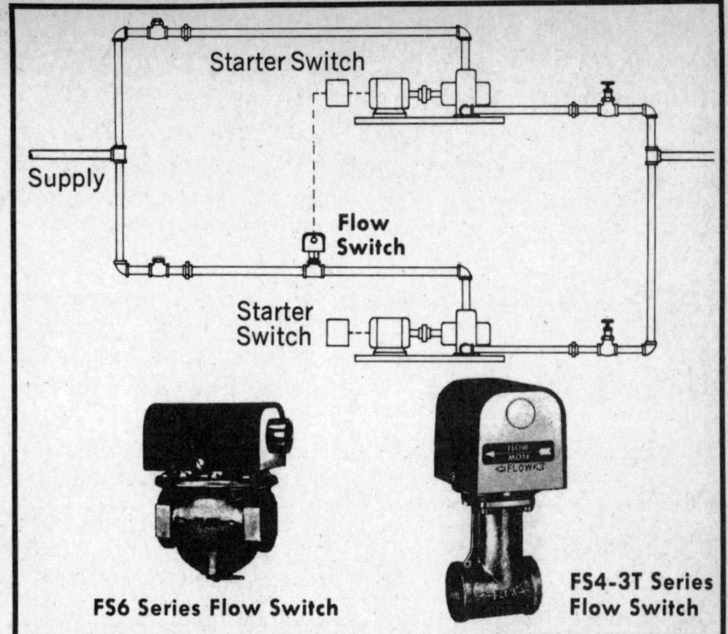

Fig. 16-7. A flow switch used to control a primary and standby pump.

restored, the temperature-sensing device will gradually restore normal circulation through the boiler, eliminating sudden temperature changes within the boiler.

Flow switches may be further used to actuate a standby pump in the event the primary circulating pump fails. Since modern hot-water heating systems are dependent upon controlled circulation, a standby circulating pump is often installed in case the primary pump fails. In the system in Fig. 16-7, the primary pump and standby pump are installed in parallel in a simple return system. A flow switch is installed in a discharge line of the primary pump and connected electrically to the starter of the standby pump.

During normal operation, the flow in the primary-pump discharge line keeps the flow switch open and the standby pump remains stopped. Should the primary pump fail, the lack of flow in the discharge line causes the contact in the flow switch to close, starting the standby pump. The flow switch may also be used to actuate a signal in a central control panel to indicate whether there is flow or no flow in the pump discharge line.

Appendix I

Manufacturers of Heating and Cooling Products

Ace-Sycamore, Inc.
448 Dekalb Ave.
Sycamore, IL 60178
Products: exhaust fans

Admiral Corp.
3800 Cortland
Chicago, IL 60647
Products: air conditioners, wall and window

Air King Corp.
3050 N. Rockwell Ave.
Chicago, IL 60618
Products: circulating fans, dehumidifiers, exhaust fans, humidifiers, range hoods, heat/fan combinations, bathroom heaters, ceiling heaters, forced-air heaters, infrared heaters, thermostats

Air King Lighting Corp.
6021 Bandini Blvd.
Los Angeles, CA 90022
Products: infrared heaters

Aitken Products, Inc.
566 N. Eagle St.
Geneva, OH 44041
Products: portable and permanent infrared heaters, downflow and unit heaters

American Air Filter Co., Inc.
215 Central Ave.
Louisville, KY 40208
Products: air conditioners, circulating fans, exhaust fans

American Dryer
1124 E. Franklin St.
Huntington, IN 46750
Products: hand dryers

American Electric Controls
250 Mill St.
Bellville, NJ 07109
Products: thermostats and controls, electronic proportional controls for electric heat, electronic controls for automatic air conditioning, air handling and proportional electric heating

Apextro Prods. Co.
1770 Workman St.
Los Angeles, CA 90031
Products: infrared heaters

Arrow-Hart, Inc.
103 Hawthorn St.
Hartford, CT 06106
Products: contactors, heating

Arvin Industries, Inc.
Div: Consumer Products
1531 13th St.
Columbus, In 47201
Products: baseboard heaters, bathroom heaters, cable/tapes, ceiling heaters, ceiling panels (heating) floor heaters, forced-air heaters, thermostats, wall panel heaters

Athena Controls, Inc.
2 Union Hill Rd.
W. Conshohocken, PA 19428
Products: controls, electric heat; thermostats

Barber-Colman Co.
P.O. Box 99
Rockford, IL 61101
Products: light/heat ceiling panels; solid-state electric-heat controls; thermostats

Berko Electric Mfg. Corp.
P.O. Box 365
Michigan City, IN 46360

Products: baseboard heaters, bathroom heaters, cable/tapes, heating; ceiling heaters, ceiling panels (heating), duct heaters; forced-air heaters, furnaces, electric, infrared heaters; thermostats; wall panel heaters; unit heaters

Brasch Mfg. Co., Inc.
11880 Dorsett Rd.
Maryland Hgts., MO 63043
Products: controls, electric heat, duct heaters, unit heaters

Broan Mfg. Co.
926 W. State St.
Hartford, WI 53027
Products: exhaust fans; range hoods; bathroom heaters, ceiling heaters, infrared heaters

Bryant Electric
1421 State St.
Bridgeport, CT 06602
Products; baseboard heaters, bathroom heaters, cable/tapes, heating, ceiling heaters, duct heaters, forced-air heaters, infrared heaters; thermostats

Cadet Mfg. Co.
6125 NE 105th
Portland, OR 97220
Products: baseboard heaters; electric fireplaces

Cam Industries, Inc.
18250 68th Ave.
So. Kent, WA 98031
Products: electric water heaters

Carrier Air Conditioning Co.
Carrier Parkway
Syracuse, NY 13201
Products: air conditioners, central unit residential; duct heaters, forced-air heaters; furnaces, electric; heat pumps

Ceilheat, Inc.
P.O. Box 10066
Knoxville, TN 37919
Products: air conditioners, central unit residential; baseboard heaters, bathroom heaters, cable/tapes, heating, ceiling heaters, duct heaters, forced-air heaters; furnaces, electric; infrared heaters; thermostats, wall panel heaters; mineral-insulated heating cable

Certain-Teed Saint Gobain
Valley Forge, PA 19481
Products: insulation, thermal

Chemelex, Inc.
837 Second Ave.
Redwood City, CA 94063
Projects: cable/tapes, heating

Chromalox Comfort Conditioning
8100 Florissant
St. Louis, MO 63136
Products: circulating fans; exhaust; range hoods; baseboard, bathroom, ceiling, duct, floor, forced-air, infrared heaters; cable/tapes, heating; furnaces, electric; thermostats; water heaters, electric

Continential Radiant Glass Htg. Corp.
215-B Central Ave.
E. Farmingdale, NY 11735
Products: baseboard, bathroom, infrared, wall panel heaters; ceiling panels (heating); controls, electric heat; radiators, electric

Control Devices, Inc.
1007 Ferry Rd.
Doylestown, PA 18901
Products: controls, electric heat; thermostats

Controls Co. of America
9655 Soreng Ave.
Schiller Pk., IL 60176
Products: thermostats

Crydom Controls
233 Kansas St.
El Segundo, CA 90245
Products: contactors, heating (solid-state); controls, electric heat (solid-state)

D & R Mfg., Inc.
5413 NE Columbin Bl.
Portland, OR 97218
Products: baseboard heaters; furnaces, electric; thermostats

Dayton Electric Mfg. Co.
5959 W. Howard St.
Chicago, IL 60648
Products: circulating fans; exhaust fans

Dunham-Bush, Inc.
175 South St.
W. Hartford, CT 06110
Products: air conditioners, central unit residential

Eagle Electric Mag. Co., Inc.
23–10 Bridge Plaza So.
L.I.C., NY 11101
Products: heater elements

Easy Heat Wirekraft
555 N. Michigan St.
Lakeville, IN 46536
Products: cable/tapes, heating; electric cable; floor heaters; infrared heaters; insulation, thermal; thermostats; water heaters, electric

Ebco Mfg. Co.
365 N. Hamilton Rd.
Columbus, OH 43213
Products: dehumidifiers; humidifers

Electric-Aire Corp.
16924 State St.
South Holland, IL 60473
Products: dryers, hand

Electric Furnace-Man
4th & Furnace St.
Emmaus, PA 18049
Products: baseboard heaters; bathroom heaters; furnaces, electric; infrared heaters

Electri-Heat Co.
Box 3646 CRS
Johnson City, TN 37601
Products: air conditioners, central unit residential; air conditioners, room console; air conditioners, wall; air purifiers; baseboard, bathroom ceiling, duct, floor, forced-air, infrared, wall panel, window heaters; cable/tapes, heating; electric cable; insulation, thermal; thermostats

Elektra Systems, Inc.
144 Marine St.
Farmingdale, NY 11735
Products: bathroom heaters (portable); cove or cornice heaters

Everwarm Dept.
P.O. Box 959
Manchester, NH 03105
Products: air conditioners, central unit residential; baseboard heaters; bathroom heaters; cable/tapes, heating; ceiling heaters, floor heaters, forced-air heaters; furnaces, electric; thermostats

Fedders Corp
Edison, NJ 08817
Products: air conditioners, central unit residential; wall and window air conditioners; dehumidifiers; humidifiers; furnaces, electric; heat pumps

Federal Pacific Electric Co.
150 Ave. L
Newark, NJ 07101
Products: air conditioners, central unit residential; baseboard, bathroom, ceiling, floor, forced-air, infrared heaters; cable heating; ceiling panels (heating); furnaces, electric; thermostats

Floorlevel Comfort Systems
10 Maryland Ave.
Baltimore, MD 21222
Products: baseboard heaters

Frank Electric Corp.
Willow Springs Lane Rd. 5
York, PA 17405
Products: control panels for electric heating

Gaylord Industries
6775 SW McEwan
Lake Oswego, OR 97034
Products: range hoods

General Electric Co.
Appliance Pk Bldg. 6
Louisville, KY 40218
Products: central air conditioners; humidifiers; electronic air cleaners

General Electric Co.
1 Progress Rd.
Shelbyville, IN 46176
Products: duct, forced-air, infrared, unit heaters; thermostats

Gibson Refigerator Sales Corp.
515 W. Gibson Dr.
Greenville, MI 48838
Products: air conditioners, window

Govern Electric Heater Corp.
5518 Avenue N.
Brooklyn, NY 11234
Products: baseboard heaters; thermostats; cove heaters

Hamilton Humidity, Inc.
3757 W. Touhy Ave.
Chicago, IL 60645
Products: humidifiers; electronic air cleaners

Heat Controller, Inc.
1900 Wellworth Ave.
Jackson, MI 49203
Products: central unit, wall, window air conditioners; electronic air cleaners; dehumidifiers; humidifiers; duct heaters; furnaces, electric; heat pumps

Home Metal Products Co.
750 Central Expressway
Plano, TX 75074
Products: exhaust fans; range hoods

Honeywell Residential Div.
2701 4th Ave. S.
Minneapolis, MN 55408
Products: electronic air cleaners; air conditioning/humidity controls; thermostats; heating equipment controls; electric heat sequencing controls

Hunter
2500 Frisco Ave.
Memphis, TN 38114
Products: circulating fans; exhaust fans; baseboard, bathroom, ceiling, floor, forced-air, infrared heaters; cable/tapes, heating; ceiling panels (heating); furnaces, electric; thermostats

Hotpoint
Appliance Pk.
Louisville, KY 40225
Products: air conditioners, wall; air conditioners, window

Hygrodynamics Facility
8030 Georgia Ave. Dept D14-2
Silver Spring, MD 20910
Products: electronic humidity controls

Intertherm, Inc.
3800 Park Ave.
St. Louis, MO 63110
Products: air conditioners, central unit residential; hot-water electric baseboard heaters; under-cabinet heaters; floor heaters; furnaces, electric; wall panel heaters

ITT
801 Allen Ave.
Glendale, CA 91201
Products: thermostats; heating controls

Infrared Corp. of America
240 Old Country Rd.
Hicksville, NY 11802
Products: bathroom heaters; ceiling heaters; infrared heaters

Johns-Manville
P.O. Box 5108
Denver, CO 80217
Products: insulation, thermal

Johnson Service Co.
507 E. Michigan St.
Milwaukee, WI 53201
Products: humidifiers; thermostats

Kalglo Electronics Co., Inc.
901 Tilghman St.
Allentown, PA 18102
Products: infrared heaters (unbreakable); solid-state modulating heat controls

Klockner Moeller Corp.
4 Strathmore Rd.
Natick, MA 01760
Products: heating contactors

Legend Electric Mfg. Corp.
140 Dupont St.
Plainview, NY 11803
Products: baseboard heaters; wall panel heaters

Loyola Industries, Inc.
1526 W. 240th St.
Harbor City, CA 90710
Products: thermostats; solid-state electric heat controls

Luminator
1200 E. Dallas North Pkwy.
Plano, TX 75074
Products: infrared heaters

Luxaire, Inc.
West of Filbert
Elyria, OH 44035
Products: duct heaters; furnaces, electric; heat pumps; air conditioners, central unit; air purifiers

Magnecraft Electric Co.
5575 N. Lynch
Chicago, IL 60630
Products: contactors, heating

Markel Electric Prods., Inc.
145 Seneca St.
Buffalo, NY 14203
Products: wall air conditioners; baseboard, bathroom, ceiling, floor, forced-air, infrared, window, unit heaters; ceiling panels (heating); thermostats

Martin Electric Heat
Elm St.
Athens, AL 35611
Products: baseboard, bathroom, ceiling, forced-air, infrared heaters; cable/tapes, heating; furnaces, electric; ceiling panels (heating); thermostats

McGraw-Edison Co.
704 N. Clark St.
Albion, MI 49224
Products: central unit, wall, window air conditioners; dehumidifiers; humidifers; baseboard, bathroom, ceiling, duct, forced-air, infrared heaters; cable/tapes, electric; furnaces, electric; thermostats

Miami-Carey Co.
203 Garver Rd.
Monroe, OH 45050
Products: exhaust fans; range hoods; bathroom, infrared heaters

Modine Mfg. Co.
1500 DeKoven Ave.
Racine, WI 53401
Products: air conditioners; unit ventilators; multizone rooftop units; cabinet unit heaters

Moody Products Co.
3014 E Woodbridge
Detroit, MI 48207
Products: infrared heaters

Mopeco/Apparatus
P.O. Box 666
Westminster, CO 80030
Products: forced air heaters

Mueller-Climatrol Corp.
2005 W. Oklahoma Ave
Milwaukee, WI 53215
Products: central unit, room console, wall, window air conditioners; circulating fans; electronic air cleaners; humidifiers; duct heaters; forced-air heaters; furnaces; heat pumps

N.A. Electric Lamp Co.
1520 N. 13th St.
St. Louis, MO 63106
Products: infrared bulbs, incandescent

Nelson Electric
P.O. Box 726
Tulsa, OK 74101
Products: cable/tapes, heating

Northwest Fdry & Frnc Co.
2345 SE Gladstone St.
Portland, OR 97202
Products: baseboard heaters; ceiling heaters; furnaces, electric

Nutone
Madison & Redbank Rds.
Cincinnati, OH 45227
Products: exhaust fans; range hoods; bathroom heaters; ceiling heaters

Oren Corp.
P.O. Box 2446
Muncie, IN 47404
Products: thermal insulation

Pal-O-Pak Insulation Co., Inc.
135 Cottonwood Ave.
Hartland, WI 53029
Products: thermal insulation

Penn Controls
2221 Camden Ct.
Oak Brook, IL 60521
Products: thermostats; proportional staging controls.

Pen Ventilator Co., Inc.
11th & Allegheny
Philadelphia, PA 19140
Products: exhaust fans

Porter Co., Inc., H.K.
Porter Bldg.
Pittsburgh, PA 15219
Products: room console air conditioners; dehumidifiers; exhaust fans; humidifiers

Radiant Electric Heat Co.
1222 S. 18 St.
Centreville, IA 52544
Products: central unit air conditioners

Singer Co., The
62 Columbus St.
Auburn, NY 13021
Products: air conditioners, central unit and wall; humidifiers; dehumidifiers; baseboard, bathroom, ceiling, duct, floor, forced-air, infrared wall panel, explosion-proof heaters; cable heating; ceiling panels (heating); electric furnaces; heat pumps; thermostats

Steward Inds., Inc.
320 E. St. Joseph
Indianapolis, IN 46202
Products: exhaust fans; range hoods

Sunwarm, Inc.
P.O. Box 266
Kingsport, TN 37662

Products: central unit air conditioners; baseboard, bathroom, ceiling, forced-air, infrared, wall heaters; heating cable & tapes; electric furnaces; thermostats

Swan Mfg. Co.
2600 W. Fourth Plain Blvd.
Vancouver, WA 98663
Products: baseboard, ceiling heaters; electric heat controls; fire places; electric thermostats

Swanson, Inc.
607 S. Washington St.
Owosso, MI 48867
Products: exhaust fans; range hoods

Teledyne Still-Man Mfg.
429 E. 164 St.
Bronx, NY 10456
Products: electric heating elements

Tennessee Plastics, Inc.
P.O. Box T CRS
Johnson City, TN 37601
Products: central unit air conditioners, electronic air cleaners; baseboard, bathroom, ceiling, duct, floor, forced-air, infrared, wall panel heaters; cable/tapes, heating; electric furnaces; thermostats

Texas Instru., Inc.
34 Forest St.
Attleboro, MA 02703
Products: thermostats

Thermador
5119 District Blvd.
Los Angeles, CA 90040
Products: exhaust fans; range hoods; hood ventilators; bathroom ceiling, forced-air, infrared, wall panel heaters; thermostats

Tranter Mfg., Inc.
735 E. Hazel
Lansing, MI 48909
Products: ceiling panels (heating and cooling)

United Elec. Controls Co.
85 School St.
Watertown, MA 02172
Products: thermostats

U.S. Mineral Prods. Co.
Furnace St.
Stanhope, NJ 07874
Products: thermal insulation

Ventrola Mfg. Co.
501 S. Chestnut St.
Owosso MI 48867
Products: exhaust fans; range hoods; bathroom, ceiling heaters

Vulcan Radiator Co., The
775 Capitol Ave.
Hartford CT 06106
Products: baseboard and duct heaters; electric radiators

Ward Leonard Elec. Co., Inc.
31 South St.
Mt. Vernon NY 10550
Products: baseboard, bathroom, ceiling, floor, forced-air heaters; thermostats

Westinghouse
P.O. Box 2510
Staunton, VA 24401
Products: central unit and wall air conditioners; humidifiers; forced-air heaters; electric furnaces; heat pumps

White-Rodgers
9797 Reavis Rd.
St. Louis, MO 63123
Products: air purifiers; electronic air cleaners; thermostats and controls

Appendix II

Glossary

British Thermal Unit (Btu): A unit for measuring quantity of heat. It is approximately the heat required to raise the temperature of a pound of water 1 degree Fahrenheit.

Btu per Hour (Btuh): A unit for measuring the rate of heat transfer.

Conduction, Thermal: Conduction is the process of heat transfer through a material in which energy is transmitted from particle to particle without displacement of the particles.

Convection: Convection is the process of heat transfer by movement of fluid. Natural convection is due to differences in density from temperature differences. Forced convection is produced by mechanical means.

Design Temperature: The design temperature is the temperature an apparatus or a system is designed to maintain (inside design temperature) or operate against (outside design temperature) under the most extreme conditions to be satisfied. The difference between the inside and the outside design temperatures is the design temperature difference.

Installed Resistance (R/): The installed resistance is the thermal resistance of insulation when applied according to the manufacturer's instructions in the building section described.

Radiation: Radiation is the process of heat transfer by the transmission of electromagnetic waves between two or more bodies in sight of each other. These radiant waves are converted to heat only after being absorbed by matter.

Temperature: Temperature is the thermal state of matter considered with reference to its power of communicating heat to other matter.

Thermal Conductivity (k): The number of heat units (Btu) that will pass through 1 square foot of a uniform material 1 inch thick in 1 hour with a 1° F difference in temperature between the two surfaces of the material.

Thermal Resistivity (r): The ability of unit thickness of a uniform material to retard or resist the flow of heat. It is the reciprocal of thermal conductivity (1/k).

Thermal Resistance (R): The ability of a material or combination of materials to retard or resist the flow of heat. It is the reciprocal of "U".

Thermal Resistance Unit (ru): The word "ru" is used as an abbreviation for resistance unit.

"U" (Overall Coefficient of Heat Transfer): The amount of heat flow, expressed in Btu per hour per square foot per degree F temperature difference between the air on the inside and the air on the outside of a building section (wall, floor, roof, or ceiling). The term is applied to combinations of materials, and also to single materials, such as window glass, and includes the surface conductance on both sides. This term is frequently called the "U-value." (For conversion, U = 3.412W.)

Vapor Barrier: A material which retards the transmission of water vapor.

"W": The U factor converted into electrical terms for calculations for electric heating. It is the amount of heat flow, expressed in watts per square foot per degree F temperature difference between air on the inside and air on the outside of the building section (wall, floor, roof, or ceiling). (For conversion, W = 0.293U.)

Watt: The rate of flow of electrical energy (not the quanitity, but the rate). One watt is equivalent to 3.413 Btuh.

Appendix III
Heating and Cooling Specifications

When an architect is commissioned to design a home for a client, he usually prepares complete working drawings and specifications which include the electrical and mechanical systems within the home. You are already familiar with drawings, as the basic procedure for making them was described earlier in this book. The specifications for a home or building are the written description of what is required by the owner, architect, or engineer in the way of materials, their quality, and the method in which they are to be installed.

For this reason, the homeowner should be familiar with typical heating and cooling specifications in order to approach his own work more intelligently; to be able to understand the construction documents if he is having the work done by others; and to help him ensure a safe, practical installation regardless of who installs his system.

The following sample illustrates typical heating and cooling specifications for a residential installation.

DIVISION 15—MECHANICAL
SECTION 15010—GENERAL PROVISIONS

Consult the index to be certain that the set of documents and specifications is complete. Report omissions or discrepancies to the Architect and/or Engineer.

1. Scope of the Work:

a. The scope of the work included under this section of the specifications shall include a complete heating, ventilating, and air-conditioning system as shown on the plans and as specified herein. The Mechanical Contractor shall provide all supervision, labor, materials, equipment, machinery, and any and all other items necessary to complete the system. The Mechanical Contractor shall note that all items of equipment are specified in the singular; however, the Contractor shall provide and install the number of items of equipment as indicated on the drawings and as required for the complete system.

b. It is the intention of the specifications and drawings to call for finished work, tested, and ready for operation. Wherever the word "provide" is used, it shall mean "provide and install complete and ready for use."

c. Any apparatus, appliance, material, or work not shown on the drawings but mentioned in the specifications, or vice versa, or any incidental accessories necessary to make the work complete and perfect in all respects and ready for operation, even if not particularly specified, shall be furnished, delivered, and installed by the Contractor without additional expense to the Owner.

d. Minor details not usually shown or specified, but necessary for proper installation and operations, shall be included in the Contractor's estimate, the same as if herein specified or shown.

e. With submission of bid, the Contractor shall give written notice to the Architect and/or Engineer of any materials or apparatus believed inadequate or unsuitable, in violation of laws, ordinances, rules, or regulations of authorities having jurisdiction; and any necessary items or work omitted.

In the absence of such written notice, it is mutually agreed the Contractor has included the cost of all required items in this proposal, and that he will be responsible for the approved satisfactory functioning of the entire system without extra compensation.

2. Mechanical Drawings:

a. The mechanical drawings are diagrammatic and indicate the general arrangement of systems and work included in contract (do not scale the drawings). Consult the architectural drawings and details for exact location of fixtures and equipment; where same are not definitely located, obtain this information from the Architect.

b. Contractor shall follow the drawings in laying out work and check drawings of other trades to verify spaces in which work will be installed. Maintain maximum headroom and space conditions at all points. Where headroom or space conditions appear inadequate, the Architect and/or Engineer shall be notified before proceeding with installation.

c. If directed by the Architect and/or Engineer, the Contractor shall, without extra charge, make reasonable modifications in the layout as needed to prevent conflict with work of other trades or for proper execution of the work.

3. Codes, Permits, & Fees:

a. Contractor shall give all necessary notice, obtain all permits, and pay all government taxes, fees, and other costs, including utility connections or extension, in connection with his work; file all necessary plans, prepare all documents and obtain all necessary approvals of the governmental departments having jurisdiction; obtain all required certificates of inspection for his work and deliver same to the Architect and/or Engineer before request for acceptance and final payment for the work.

b. Contractor shall include in the work, without extra cost to the Owner, any labor, materials, services, apparatus, drawings (in addition to contract drawings and documents) in order to comply with all applicable laws, ordinances, rules, and regulations, whether or not shown on drawings and/or specified.

c. All materials furnished and all work installed shall comply with the rules and recommendations of the National Board of Fire Underwriters, with the requirements of local utility companies, with the recommendations of the Fire Insurance Rating

Organization having jurisdiction, and with the requirements of all governmental departments having jurisdiction.

4. Shop Drawings:

a. The Mechanical Contractor shall submit five (5) copies of the shop drawings to the Architect and/or Engineer for approval within thirty (30) days after the award of the general contract with the exception of temperature controls or such items which require approved shop drawings on other equipment before the item can be selected. Such items shall be submitted within sixty (60) days. If such a schedule cannot be met, the Mechanical Contractor may request in writing an extension of time from the Architect. If the Mechanical Contractor does not submit shop drawings in the prescribed time, the Architect has the right to select the equipment.

b. Shop drawings shall be submitted on all major pieces of heating, ventilating, and air conditioning equipment. Each item of equipment proposed shall be a standard catalog product of an established manufacturer. The shop drawings shall give complete information on the proposed equipment such as: capacity, size, construction, material, dimensions, arrangement, operating clearances, performance characteristics, weight, and rating authority. Each item of the shop drawing shall be properly labeled, indicating the intended service of the material, the job name, and Mechanical Contractor's name.

c. The Mechanical Contractor, before submitting the shop drawings of the equipment to the Architect and/or Engineer, shall check each item of the shop drawings to verify the appropriateness of the equipment. The check shall include such considerations as: will the equipment physically fit into the allotted space; is the equipment proper for the job; does the voltage match that of the electric service; are provisions for connections proper; are code requirements met?

d. The shop drawings shall be neatly bound in five (5) sets and submitted to the Architect and/or Engineer with a letter of transmittal. The letter of transmittal shall list each item submitted along with the manufacturer's name.

e. Approval rendered on shop drawings shall not be considered as a guarantee of measurements or building conditions. Where drawings are approved, said approval does not mean that drawings have been checked in detail, and said approval does not in any way relieve the Contractor from his responsibility, or necessity of furnishing material or performing the work as required by the contract drawings and specifications.

5. Cooperation with Other Trades:

a. The Mechanical Contractor shall give full cooperation to other trades and shall furnish in writing, (with copies to the Architect and/or Engineer) any information necessary to permit the work of all trades to be installed satisfactorily and with least possible interference or delay.

b. Where the work of the Mechanical Contractor will be installed in close proximity to work of other trades, or where there is evidence that the work of the Mechanical Contractor will interfere with the work of other trades, he shall assist in working out space conditions to make a satisfactory adjustment. If so directed by the Architect and/or Engineer, the Mechanical Contractor shall prepare composite working drawings and sections at a suitable scale clearly showing how his work is to be installed in relation to the work of other trades. If the Mechanical Contractor installs his work before coordinating with other trades or so as to cause any interference with work of other trades, he shall make changes in his work necessary to correct the conditions without extra charge.

6. Equipment Deviation:

a. PROTOTYPE means the manufacturer's equipment or material which the job was designed around. ACCEPTABLE means the manufacturer may substitute equipment or materials provided the quality is fully equal to the prototype and the equipment or materials meet the specifications. APPROVED EQUAL means the manufacturer's equipment or materials shall be approved by the Architect and/or Engineer in advance of the contractor's proposal as being fully equal in quality to the prototype and meeting the specifications.

b. Where deviating from prototype equipment and materials requires additional expense, such as change in structural design, ductwork, piping, electrical wiring, etc., the Contractor shall bear the expense of such changes without additional expense to the Owner.

7. Electrical Wiring:

a. The Mechanical Contractor shall, regardless of voltage, furnish and install all temperature control wiring and all interlock wiring and equipment control wiring for the equipment that the Mechanical Contractor furnishes. The Electrical Contractor will furnish and install power wiring to the mechanical equipment and make all electrical connections unless otherwise noted on the drawings.

b. All electrical wiring furnished under the mechanical contract shall conform with the Electrical Specifications.

8. Electric Motors:

a. The Mechanical Contractor shall provide and install all electric motors for his equipment unless otherwise noted. All motors shall be NEMA standard design for quiet operation. The motors shall be ample in size to operate at their proper load and full speed continuously without causing noise, vibration, or temperature rise in excess of the rating.

b. Motors with belted drives shall be mounted in a manner permitting belt adjustment. All belts shall be adjusted before turning the project over to Owner.

9. Electric Motor Starters:

a. The Mechanical Contractor shall furnish all motor starters with accessories as required, such as start-stop push-button switches, hand-off auto selector switches, pilot lights, remote switches, auxiliaries, contacts, and overload thermal units or heaters for his part of the work to the Electrical Contractor for installing, unless otherwise noted.

b. All three-phase motors shall be supplied with a magnetic line-voltage starter unless otherwise noted, with low-voltage protection on each phase. All single-phase motors shall have manual starters, unless otherwise noted, with overload protection.

c. All starters shall be horsepower-rated, with the rating not less than the motor horsepower. The overload thermal unit or heater shall be selected from the full-load amperes as listed on the nameplate of the motor to be protected. The heater rating shall not exceed 115% of motor full load amperes.

d. Accepted Manufacturers—GENERAL ELECTRIC, WESTINGHOUSE, SQUARE "D," ALLEN BRADLEY, CLARK, CUTLER-HAMMER.

10. Operating Instructions:

a. The Mechanical Contractor shall instruct the Owner or his representative in the operation and maintenance of the system. The instructions shall not terminate until the Owner or his representative is completely familiar with the system.

b. All the manufacturer's installation, operating, and maintenance manuals that accompany the equipment shall be securely fastened to the equipment or turned over to the Owner or his representative.

11. Before Acceptance:

a. All equipment shall be cleaned and left in a neat manner. Any paint scratches shall be "touched-up" with factory-color paint.

12. Performance Test:

a. The system shall be tested at the completion of the building. It shall be established that all units are producing to their designed capability and all controls are performing satisfactorily. The system shall be checked for vibration and excessive noise and all such conditions corrected.

13. As-Built Drawings:

a. The Mechanical Contractor shall keep accurate records of all deviations in work as actually installed from work indicated on the drawings. When work is completed, two complete sets of marked-up prints shall be delivered to the Architect.

14. Separate Contracts:

a. In the event that the mechanical section of the drawings and specifications are covered by more than one contractor (such as a plumbing contractor and a heating contractor), it shall be the responsibility of the separate contractors to clearly state in their proposal the items covered by their proposal. Any such misunderstandings deriving from separate contracts shall be mediated by the General Contractor.

Section 15050—Basic Materials and Methods

Consult the index to be certain that the set of documents and specifications is complete. Report omissions or discrepancies to the Architect and/or Engineer.

1. Ductwork Material and Workmanship

a. GENERAL: The sheet metal contractor shall provide and install a complete first-class ductwork system including connections to all equipment. The sheet metal contractor shall coordinate his work with other trades. The measurements for fabricating the duct shall be made on the job and not scaled from the drawings.

b. CONSTRUCTION: All ductwork shall be constructed of galvanized sheet metal, unless otherwise noted, using the sheet-metal gauge reinforcing and joints as recommended by the latest edition of the ASHRAE Guide and Data Book for low-pressure ducts. The gauge of sheet metal shall be stamped on both sides of the sheet metal. All seams shall be on the outside of the duct. The sizes of the ducts are shown on the drawings. In case the size is not shown, the size shall be requested from the Architect. In difficult areas or where required, the shape of the duct may be changed when approved by the Architect and as long as the cross-sectional area of the duct is maintained. Where a duct liner is used, the size of the duct shall be the inside free area. All duct transitions shall be gradual. All elbows shall be made with a minimum inside radius equal to one-half the width of the duct. Any elbows requiring a radius below the minimum shall be made square using turning vanes. Turning vanes shall be Tuttle and Bailey or equal.

c. INSTALLATION: The ductwork shall be erected in a neat and workmanlike manner. The ductwork shall be hung or supported from the building structure in a manner such as to prevent vibration. Galvanized sheet-metal straps or galvanized support rods shall be used to support the duct rigidly.

d. MANUAL AND SPLITTER DAMPERS: Shall be installed as shown on the drawings and required by the National Fire Prevention Association pamphlet 90A or the local fire code.

e. FLEXIBLE CONNECTION: Shall be provided in the ductwork connected to the fans. At least 1 inch of slack shall be allowed in all flexible connections to insure that no vibration is transmitted. The duct transition to and from fans shall be made gradual to prevent turbulence. All flexible connections shall be fireproof woven asbestos.

f. AIR BALANCE: The entire air system shall be balanced to the satisfaction of the Architect. The work includes adjusting fan speeds, manual and splitter dampers, grilles, registers, and ceiling diffusers to obtain the desired air distribution.

g. GRILLES, REGISTERS & CEILING DIFFUSERS:

(1) Grilles, registers, and ceiling diffusers shall be provided as shown on the drawings and installed in accordance with the

manufacturer's recommendations. Dampers shall be installed on this equipment where shown and where required to balance the air system. All accessories shall be manufactured by U.S. REGISTER, CARNES, or KROEGER unless otherwise noted.

(2) Before locating grilles, registers, and ceiling diffusers, check the architectural and electrical drawings to make sure that there is no conflict with floor moldings, electrical outlets, lighting fixtures, or any other obstruction. Low sidewall grilles and registers (LSWR) shall be mounted with the bottom edge six to eight inches above the floor with the vanes turned down. High sidewall grilles and registers (HSWR) shall be mounted next to the ceiling or as shown on the architectural drawings.

(3) Air extractors shall be provided and installed as shown on the drawings. Provision shall be made to adjust the air extractor from the exterior of the ductwork. When an air extractor is installed, no damper for the register is required.

h. LOUVERS:

(1) Supply air, exhaust air, and combustion air louvers shall be installed as shown on the drawings or as required. The louvers shall be furnished by the Mechanical Contractor unless specified under the architectural section. The louvers shall be provided with 1/2-inch aluminum bird screen and duct collars where required and installed in a manner such that no water will enter the building.

2. Insulation Material and Workmanship:

a. GENERAL: Insulation shall be applied to all parts of the system as listed in other sections of the specifications or on the drawings. The insulating material shall be manufactured by OWENS-CORNING, GUSTIN-BACON, ARMSTRONG, OR

JOHNS MANVILLE or approved equal. All insulating materials shall be installed in accordance with the manufacturer's recommendation and shall be installed in a neat and workmanlike manner by personnnel regularly employed in the pipe-covering trade. Insulation shall be applied to all parts of the system as specified in the appropriate section of the specifications. All pipe and duct insulation, covering, and lining shall have a flame-spread rating of not over 25; smoke developed, 50; and fuel contributed, 50.

b. PIPE INSULATION:

(1) *Fiberglass Pipe Insulation:*

(a) Shall be a pipe covering with white flame-retardant vapor-barrier jacket (W.F.R.J.) unless otherwise noted. Exposed piping shall have a 7 1/4-pound density and concealed piping shall have a 4-pound density. Pipe insulation under saddles shall be 7 1/4-pound density.

(b) Pipe insulation shall be applied by drawing the jacket tight and smooth. Secure longitudinal joints with self-sealing lap or fire-retardant vapor-barrier adhesive, Tightly cover the end joints with 3-inch-wide strips of identical jacket material secured in the same manner. Any puncture to the vapor barrier must be completely sealed with flame-retardant mastic.

(c) Specialities such as valves, fittings, flanges, etc., shall be insulated with molded or fabricated insulation of thickness equivalent to and composition identical to that used on adjacent pipes. Apply two coats of mastic with a layer of open-weave glass cloth embedded in the first coat. Draw the fabric tight and smooth with a 2-inch overlap at all joints.

(d) The insulation contractor shall provide and install sheet-metal saddles for the hangers on the horizontal hot-water pipe. The saddles shall be 16-gauge sheet metal and 8-inches long.

(2) *Flexible Pipe Insulation:* Shall be foamed-plastic insulation. Insulation can be slipped over the pipe before installation or snapped over piping already connected. All joints shall be sealed with O-C 500 adhesive. Insulation installed exposed to the weather or earth shall be protected with Benjamin Foster's Lagtone, Insul-Coustic 110 Sure Kote or Alkyd-Chlorinated Rubber Paint.

c. DUCT INSULATION:

(1) *Fiberglass Blanket:* Shall be fiberglass duct wrap insulation with factory-applied reinforced-foil flame-resistant kraft paper (RFK) vapor-barrier jacket. Wrap the insulation tightly and smoothly around the duct. On the bottom of rectangular ducts over 24 inches in width, additionally secure the insulation with mechanical clips on maximum 18-inch centers. No sagging will be permitted. Butt all edges of the insulation firmly together, lap the vapor-barrier jacket a minimum of 2 inches. Secure the laps with flame-retardant adhesive. Seal all joints and punctures with flame-retardant mastic.

(2) *Rigid-Board Duct Insulation:* Shall be fiberglass vapor-seal duct insulation with a foil-faced (FF) vapor barrier. Secure the insulation with mechanical clips on maximum 18-inch centers. Use a minimum of two rows of fasteners per side. Butt all edges of the insulation. Carefully fit the boards to ensure tight joints. All joints, seams, and punctures shall be sealed with flame-retardant vapor-barrier mastic.

(3) *Duct Liner:* Shall be fiberglass Aeroflex duct liner with smooth black-coated surface, 3-pound density, and 1-inch thickness. Adhere

the liner with fire-retardant adhesive, minimum 50% coverage. Use mechanical fasteners on maximum 16-inch centers on the top when the width exceeds 12 inches and on the sides when the height exceeds 24 inches.

d. EQUIPMENT INSULATION: Shall be fiberglass industrial insulation with standard face and 1 1/2-inch thickness, cut to fit the contour of the equipment. Hold the material in place with 1/2-inch-wide galvanized steel bands, spaced 9 inches on center. Seal all joints, punctures, and voids with vapor-barrier mastic. Apply insulating cement over 1-inch galvanized wire netting. The cement must be of a thickness sufficient to completely cover the netting in a manner such as to provide a smooth, neat surface. All removable covers and heads shall be insulated in a manner such as to provide accessibility and replacement without damage to the insulation.

3. Cutting and Patching:

a. On new work the Mechanical Contractor shall furnish sketches to the General Contractor showing the locations and sizes of all openings and chases, and furnish and locate all sleeves in inserts required for the installation of the mechanical work before the walls, floors, and roof are built. The Mechanical Contractor shall be responsible for the cost of cutting and patching where any mechanical items were not installed or were incorrectly sized or located. The Contractor shall do all drilling required for the installation of his hangers.

b. On alterations and additions to existing projects, the Mechanical Contractor shall be responsible for the cost of all cutting and patching, unless otherwise noted.

c. No structural members shall be cut without the approval of the Architect, and all such cutting shall be done in a manner directed by him. All patching shall be performed in a neat and workmanlike manner.

4. Excavation and Backfilling:

a. The Mechanical Contractor shall be responsible for excavation, backfill, tamping, shoring, bracing, pumping, street cuts, repairing of finished surfaces, and all protection for safety of persons and property as required for installing a complete mechanical system. All excavation and backfill shall conform to the architectural section of the specifications.

b. Excavation shall be made in a manner to provide a uniform bearing for pipes. The pipe elevation shall be determined by the Mechanical Contractor to meet the plumbing codes. Where rock is encountered, excavate 3 inches below pipe grade and fill with gravel to grade.

c. After required test and inspections, backfill the ditch and tamp. The first foot above the pipe shall be hand-backfilled with rock-free clean earth. The backfill in the ditches on the exterior and interior of the building shall be tamped to 90%. The Mechanical Contractor shall be responsible for any ditches that go down.

5. Equipment and Installation Workmanship:

a. All equipment shall be new and shall bear the manufacturer's name and trade name. The equipment to be furnished under each section of the specifications shall be essentially the standard product of a manufacturer regularly engaged in the production of the required type of equipment and shall be the manufacturer's latest approved design.

b. The Mechanical Contractor shall receive and store the equipment pertaining to the mechanical work properly. The equipment shall be tightly covered and protected against dirt, water, chemical or mechanical injury, and theft. The manufacturer's directions shall be followed completely in the delivery, storage, protection, and installation of all equipment and materials.

c. The Mechanical Contractor shall provide and install all items necessary for the complete installation of the equipment or required by code without additional cost to the Owner, regardless of whether the items are shown on the plans or covered in the specifications. Such items could be, but are not limited to, concrete pads, supports, vibration eliminators, additional piping and valves, motor controllers, relief-valve piping, insulation, electrical wiring, lubrication, refrigerants, start-up and service, etc. All equipment shall be installed in a neat and workmanlike manner.

d. It shall bc thc responsibility of the Mechanical Contractor to clean the equipment, make necessary adjustments, and place the equipment in operation before turning the equipment over to the Owner. Any paint that was scratched during construction shall be "touched-up" with factory-color paint to the satisfaction of the Architect. Any items that were damaged during construction shall be replaced.

6. Access Doors:

a. The Mechanical Contractor shall furnish to the lather the access doors as shown on the drawings or required for access to valves, etc. The doors shall be 12 inches square, unless otherwise noted, hinged, with metal frames. Door and frame shall be not lighter than 16-gauge sheet steel. The access door shall be of the flush type with screwdriver latching devices. The frame shall be constructed so that it can be secured to building material as required. The access doors shall be Milcor or equal.

7. Concrete Pads, Supports, and Piers:

a. The Mechanical Contractor shall be responsible for all concrete pads, supports, piers, bases, and foundations required for the mechanical equipment and piping. The concrete pads for the mechanical equipment shall be six (6) inches larger all around than the base of the equipment and a minimum of four (4) inches thick.

8. Waterproofing:

a. The Mechanical Contractor shall provide all flashing, caulking, and sleeves required where his items pass through the outside walls and roof. The waterproofing of the openings shall be made absolutely watertight.

SECTION 15650—REFRIGERATION

Consult the index to be certain that the set of documents and specifications is complete. Report omissions or discrepancies to the Architect.

1. Refrigerant Piping:

a. GENERAL: The refrigerant piping systems shall be provided complete and installed in accordance with the manufacturer's recommendations. The size of the refrigerant pipes shall be obtained from the equipment manufacturer.

b. PIPING: Shall be hard copper Type "L." All piping joints shall be made with silver solder or Silfos (copper to copper). The piping shall be charged with dry nitrogen while sweating joints.

c. PRESSURE TEST: After the refrigerant piping has been completed, the refrigerant system shall be pressure tested at a pressure of 300 PSI on the high side and 150 PSI on the low side. While the system is being pressure tested, an electronic leak detector shall be used to check for leaks after a soap bubble search. The above pressures shall be maintained on the system for a minimum of 12 hours.

d. EVACUATION: A minimum of 2000 microns shall be pulled on the system and maintained for 12 hours. The system shall be evacuated when the surrounding ambient air is not less than 60 degrees F. If the temperature is less, auxiliary heat shall be provided to insure proper evacuation conditions. The vacuum pump displacement shall be a minimum of 2 CFM for up to 15 tons. Over 15 tons a 5 CFM pump shall be used. The vacuum shall be checked with an electronic gauge.

e. MISCELLANEOUS PIPING ACCESSORIES:

(1) A removable cartridge-type combination dryer-strainer shall be installed in the liquid line. The cartridge shall be removable without disrupting any piping.

(2) A three-valve bypass shall be provided for the dryer-strainer.

(3) A sight glass shall be installed in liquid line.

(4) Shut-off valves shall be installed on the air-cooled condenser liquid line and hot gas lines.

(5) No screw fittings will be permitted, as all fittings shall be used.

SECTION 15670—AIR-CONDITIONING SYSTEM

Consult the index to be certain that the set of documents and specifications is complete. Report omissions or discrepancies to the Architect.

1. Split-System Condensing Units:

a. GENERAL: Units shall have all components assembled on one common base. These shall include: compressor, condenser coil, fan, motor, refrigerant reservoir, and operating controls. Units shall comply with ARI standard 210 and match the manufacturer and model number indicated on the drawings.

b. CASING: Casing shall be constructed of 18-gauge zinc-coated steel finished in baked enamel. Provide removable end panels for access to controls and drain holes for rain elimination. Mounting pads shall be included.

c. COMPRESSOR & CONTROLS: Compressor shall be a welded hermit-shell-type with built-in overloads. Controls shall include compressor and condenser fan contactors, low and high pressure controls, and a reset relay to prevent recycling on any safety control. A suction-line accumulator shall be provided to break up liquid slugs at abnormal conditions.

d. CONDENSER: Condenser fan shall be a direct-drive aluminum-propeller type with a permanently lubricated motor with thermal overload protection. The condenser coil shall be of wrap-around design and shall be provided with a liquid accumulator and liquid subcooler to prevent flashings. Fins shall be aluminum, mechanically bonded to copper tubes. The coil shall be provided with a protective guard.

e. EVAPORATOR COMPONENTS:

(1) *Evaporator Coils:* Shall match the condensing unit and shall be downflow, upflow, or horizontal as indicated. The tubes shall be of seamless copper with mechanically bonded aluminum fins. Coil selections shall include a drain pan and insulated casing. Units shall be as shown on the drawings.

SECTION 15800—AIR DISTRIBUTION SYSTEM

Consult the index to be certain that the set of documents and specifications is complete. Report omissions or discrepancies to the Architect.

1. Ductwork:

a. GENERAL: The sheet metal contractor shall install a complete galvanized sheet-metal duct system including manual, splitter, and automatic dampers, fire dampers, flexible connection to all fans, louvers, air intakes and exhausts, and all items required for a complete system. The system shall be installed to meet the requirements of the prevailing code and Section 15B.

2. Duct Insulation:

a. CONCEALED SUPPLY AIR AND OUTSIDE AIR DUCTS: Insulated with 1 1/2-inch fiberglass blanket duct insulation.

b. EXPOSED SUPPLY AIR DUCT: Insulate with one-inch duct liner.

c. RETURN AIR AND TRANSFER AIR DUCT: Insulate with one-inch duct liner.

3. Grilles, Registers, Ceiling Diffusers, and Louvers:

Provide grilles, registers, diffusers, and louvers as shown on the plans.

SECTION 15801—INCREMENTAL AIR CONDITIONING

Consult the index to be certain that the set of documents and specifications is complete. Report omissions or discrepancies to the Architect.

1. General Description:

a. Furnish permanently installed packaged-terminal air conditioners of sizes and capacities shown on the drawings. Wall openings, electrical circuits, and the like have been designed to fit Types EK Incremental Conditioners. The Contractor may submit similar equipment of other manufacture that satisfies these specifications, provided that, ten days prior to bid date, he submits detailed drawings for the correlation of other trades and also provided that he includes with his bids all additional costs accruing to the other trades. Conditioners shall not exceed the following dimensions unless otherwise approved: overall depth 18 3/4 inches; room cabinet height 25 inches, width 48 inches, depth 7 1/4 inches (optional depths available in 1-inch increments from 6 1/4 inches to 18 1/4 inches). Each conditioner shall consist of a wall box, outside air louver, heater section with controls, cooling chassis, and room cabinet.

2. Wall Box:

a. Shall be fabricated of 16-gauge zinc-coated phosphatized steel, enrobed in a continuous film of thermosetting plastic (epichlorhydrin and bis-phenol). It shall be in one piece, without concealed joints, and designed so the outside air louver may be fastened to the wall box from within the building.

3. Outside Air Louver:

a. Provide with each conditioner one anodized-aluminum extruded (or plate type) outside-air louver in natural (or other) finish. (Louvers furnished by others must be approved as to free area and design by the equipment manufacturer. If the louver is part of panel wall construction, it should be omitted from these specifications.)

4. Heater Section:

a. Shall contain low-density electric resistance-heating elements with safety devices to turn off heaters if for any reason the heater temperature becomes excessive. The heater section shall also contain two double-inlet centrifugal blowers with 6-inch aluminum wheels, direct-connected to a PSC motor with built-in overload protection. The motor is automatically to operate at 1050 rpm on cooling; 790 rpm on heating. The blower assembly shall be easily removable independent of the heating coils or cooling chassis.

5. Controls:

a. Provide with each heater section built-in controls as follows: an easily adjustable thermostat to regulate room air temperature through control of the refrigeration compressor on cooling, and the electric heating elements on heating. Also provide a master control of the push-button type for selecting OFF, COOL, or HEAT (OFF-ON, in the case of automatic-changeover units).

6. Cooling Chassis:

a. Shall consist of a self-contained welded hermetically-sealed air-cooled refrigeration system exclusive of evaporator fans and controls but with a cord and 5-prong plug for connecting to heat section. The compressor shall be internally spring-mounted, with vibration insulators, and shall include a PSC motor, overload protection, and a fused-type capacitor. All sheet-metal parts shall be zinc-coated, phosphatized, and enrobed in a continuous film of thermosetting plastic (epichlorhydrin and bis-phenol). Provision shall be made for easy removal and insertion of this chassis as a unit without interrupting the heating function and without the use of tools. Stainless-steel fasteners shall be used throughout. The direct-connected condenser blower assembly shall be of centrifugal type with a 7 1/2-inch DWDI aluminum wheel, designed to re-evaporate all condensate on the condenser surface without drip, splash, or spray on the building exterior, and to introduce ventilating air at a minimum pressure of 0.6 inches on a water gauge. Provide an automatic motor-operated damper for ventilation air with a concealed overriding switch. The refrigerant metering device shall consist of a capillary restrictor supplemented by a constant-pressure expansion valve, designed so as to prevent frosting of the evaporator coil and short cycling of the compressor at outdoor temperatures down to 35°F, while providing not less than 100% of rated cooling capacity. The cooling chassis shall also include an easily removable, washable air filter, arranged to filter both ventilating and recirculated air.

7. Room Cabinet:

a. Shall be 18-gauge furniture steel finished in mist grey baked enamel. Adjustable kick plates shall be provided. The front panel shall be tamper-proof and removable without the use of tools to provide full access to filters and the cooling chassis. Discharge grilles shall be 4-way adjustable. Return air shall enter between the kick plate and front panel.

8. Service:

a. Include in the bid and state who will be responsible for adjusting and starting the conditioners, demonstrating their proper operation to the Owner or his representative, and rendering necessary maintenance service (other than filter cleaning) during the first year of operation.

Index